THE CONSUMER'S
GOOD CHEMICAL GUIDE

THE CONSUMER'S GOOD CHEMICAL GUIDE

A jargon-free guide to the chemicals of everyday life

JOHN EMSLEY

Science Writer in Residence
Imperial College of Science, Technology & Medicine, London

W.H. FREEMAN
SPEKTRUM

OXFORD · NEW YORK · HEIDELBERG

W.H. Freeman and Company Limited
20 Beaumont Street, Oxford, OX1 2NQ
41 Madison Avenue, New York, NY 10010

British Library Cataloguing-in-Publication Data
A catalogue record for this book is available from the British Library.

Library of Congress Cataloging-in-Publication Data

Emsley, John.
The consumer's good chemical guide : a jargon-free guide to
the chemicals of everyday life / by John Emsley.
p.   cm.
Includes bibliographical references and index.
ISBN 0-7167-3034-0 (pbk)
1. Toxicology—Popular works. 2. Medical misconceptions.
I. Title.
RA1213.E47   1994   615.9—dc20   94–8286   CIP

Set by KEYWORD Publishing Services, London
Printed by The Maple-Vail Book Manufacturing Group

ACKNOWLEDGEMENTS

There are many people who helped me with information and advice during the writing of this book, and to whom I am most grateful. They are Dr Tom Addiscott, Dr John Baldwin, Dr Ian Campbell, Dr Dick Challand, Dr Bernard Dixon, Dr Peter Fell, Dr Keith Goulding, Dr Christopher Kenyon, Dr Robyn Marsack, Dr Gregory Place, Dr Ray Richards, Mr John Russell, Dr Marshall Smalley and Dr Charles Sell. For any scientific errors that still remain I take full responsibility.

My thanks also to those who helped me with their encouragement and support: my wife Joan; my children Julian and Helen; and my friends Tom Wilkie, Steven and Rosemary Ley, and Marcus and Karen Chown.

Last but not least I should like to thank my publisher Michael Rodgers, and editor Jane Gregory.

CONTENTS

9 Carbon dioxide, CO_2

Carbon dioxide, CO_2; greenhouse gases; the global CO_2 budget; global warming; the good side of carbon dioxide

INTRODUCTION

The Consumer's Good Chemical Guide is mainly a book about chemicals that are popularly regarded as anything but good; indeed many people think most of them should be banned. If you are of that opinion I hope I can persuade you to change your mind. In this book I will discuss some of the chemicals which I believe cause a lot of unnecessary alarm, but which no longer deserve to be thought of as dangerous. Many chemicals seem to come into this category, but I have chosen to discuss those which I feel cause most worry. There may be some chemicals that you are worried about which are missing from this book, but they may well be missing because they really are dangerous, like CFCs, lead or radon.

Those chemicals I have chosen to cover are those which I feel have been particularly subjected to a lot of misinformation. Sugar, artificial sweeteners, painkillers, cholesterol, animal fats, PVC, dioxins, nitrates and carbon dioxide come into this category. I have also put in two chapters on things which are entirely chemical, but you probably never think of them as such: perfume and alcohol. In the perfume chapter I will show you how my fellow chemists are trying to make the world a better place, and at the same time saving wildlife. In the chapter on alcohol I will show how a little understanding of the chemistry of this chemical can improve the quality of your life, and maybe even prolong it.

Although this book is all about chemicals, I have tried to keep it free of chemical jargon and chemical formulas. However, for those readers who want to know more about the chemistry of the materials involved, there is an appendix at the end where you can find this information.

To a large measure almost all the positive associations of the word 'chemical' have now been lost under a barrage of bad publicity, and it is increasingly difficult even for a scientist to use the word in a neutral way, simply to mean a chemical element or chemical compound. Many readers will know that to describe a product as 'free of chemicals' in meaningless, but many people take these words to mean 'free of additives'. Where possible I will try and use less emotive words like substance, compound or molecule. Yet the negative implications of 'chemical' are difficult to shake off, and phrases like 'chemical disaster' and 'chemical warfare' only serve to reinforce its poor image. Even the words 'chemical industry' now seem menacing to some people, and you may think that they deserve to be thought of in this way because of the dreadful mistakes the chemical industry has made in the past.

If we go along with the current climate of opinion, and disapprove of all things 'chemical' and blame chemists for the ills of the world, then we are in danger of undermining an area of science that can solve many of the problems we face: finding cures for cancer, Alzheimer's and AIDS; developing renewable sources of energy; replacing depleted natural resources of fossil fuels and minerals; and even lifting the threat of pollution, most of which, you may be surprised to learn, is not caused by the chemicals industry.

Here, then, is my guidebook to some parts of chemistry that impinge on all our lives. You may find some of it rather surprising, and I hope that it will change the way you see substances that are often branded as 'chemicals'. At the end of the book is a Bibliography which suggests other books that you might want to consult for further reading, and lists the sources that I have consulted during the writing of *The Consumer's Good Chemical Guide*.

Weights, volumes, areas, temperature and energy

The measurements for these quantities in *The Consumer's Good Chemical Guide* are in the metric system, which is standard throughout Europe and Australia, mainly used in the UK, but little used in North America. For the non-scientist the relative quantities may be unfamiliar. Here are the more commonly used measurements.

Weights

A tonne is 1000 kilograms (kg); a kilogram is 1000 grams (g); a gram is 1000 milligrams (mg); and a microgram is one millionth of a gram.
An Imperial ton of 2240 pounds equals 1016 kg, while a US ton of 2000 pounds equals 907 kg.
An ounce is 28 g, and a pound is 453 g. A milligram is 0.000035 ounce.

Volumes

A cubic metre (m^3) is 1000 litres (l); a litre is 1000 millilitres (ml).
The Imperial pint is equal to 568 ml, while a US pint is 473 ml.
An Imperial gallon equals 4.55 l, while a US gallon equals 3.79 l.
A cubic yard equals 0.765 m^3.

Lengths and areas

A kilometre (km) is 1000 metres (m); a metre is 100 centimetres (cm) or 1000 millimetres (mm). A hectare is 100 metres square, i.e. 10 000 m^2.
A foot equals 30.5 cm; a yard equals 0.914 m.
An acre equals 0.404 hectare.

Temperature

On the Celsius scale (°C) water freezes at 0 °C and boils at 100 °C. Body heat is 37 °C.
On the Fahrenheit scale (°F) water freezes at 32°F and boils at 212°F. Body heat is 98.6°F.

Energy

A calorie is the energy required to raise the temperature of one gram of water by one degree Celsius. A Calorie is 1000 calories, and this is the commonly used unit of energy when talking about food. It is more correctly referred to as a kilocalorie (kcal).

CHAPTER ONE

Perfume

Chanel N° 5; the chemistry of smell; making fragrant molecules; human scents; aphrodisiacs; aromatherapy

MOST of this book is about chemicals that are commonly regarded as threatening. I hope I will be able to persuade you that some of them are not—that they are good rather than bad. But to begin with I want to talk about a group of chemicals towards which you may already have very positive feelings: the fragrance chemicals. Indeed, you may not even think of them as chemicals at all; nevertheless, they are just as chemical as the other materials I will be dealing with in later chapters. Not only do fragrance chemicals give a great deal of innocent pleasure, but their synthetic versions have even been responsible for reducing the large-scale slaughter of some equally innocent animals.

The fragrances in a bottle of perfume are just chemicals with attractive aromas. Some may have originally come from exotic tropical plants, and some from exotic wild animals, but today most of them come from the mundane world of the chemistry laboratory. Perhaps we expect the fragrances that we encounter in detergents, fabric softeners and toilet cleaners to be produced by chemists, since these are merely added to mask the smell of soap or detergent. Similarly you might not be surprised to discover that cleaning aids and air fresheners for use in kitchens, living rooms and bathrooms use equally synthetic chemicals. But as far as the chemist is concerned, there is no difference between their use in such everyday products and their use in expensive perfumes.

Nevertheless, there is a difference. The melodies of the fabric softener and the air freshener are the pop songs of fragrance chemistry, whereas the symphonies of perfume are a much richer harmony, among which there may even be a few discordant notes, deliberately introduced to contrast with the harmonic ones. This is certainly true of even the world's most famous perfume, *Chanel N° 5*.

Chanel N° 5

The screen icon Marilyn Monroe was once asked what she wore in bed, and she scandalised cinema audiences in the 1950s by replying: "Only *Chanel N° 5*". When she made that remark, the perfume she referred to had been on the market for 30 years, and 40 years later it is still selling well. In the 1990s it now has to compete with over 400 other perfumes for women, and every year there are more and more of them. *Chanel N° 5* has a special place in the history of chemistry, because it made a synthetic ingredient acceptable for the first time, although most of its components were still derived from natural products. It was launched in 1921 by the Paris designer Gabrielle 'Coco' Chanel, and was created for her by Ernest Beaux to complement her collection of clothes, whose themes were elegance and simplicity. Even the bottle she chose for *Chanel N° 5* was part of this image. Little did she realise that she had launched a perfume that would immortalise her name.

Chanel N° 5 was a landmark in fragrance chemistry for two reasons. Firstly Ernest Beaux, *le grand nez*[1] who 'composed' it, chose an oil derived from the flowers of the ylang-ylang tree which grows in Philippines, to provide the middle note of the perfume. Secondly he used a purely artificial fragrance material, called **2-methylunde-canal**,[2] for the top note. This compound is one of a class of molecules

---

[1] *Le grand nez* literally means 'the big nose' and refers to those men and women who design a new perfume.

[2] Where the name of a substance is given in **bold** type, it means that there is more information about its chemistry in the Appendix.

called aldehydes, which generally have a pleasant smell, and several are now used to provide the top notes of perfumes.

The top note is the most volatile part of a perfume, and the one we detect first. The middle notes take a little longer to register, and these are often derived from flowers that give off a heavy, almost over-powering aroma, such as jasmin, tuberose, lily of the valley, lilac, carnation, rose and ylang-ylang. The base notes of a perfume are the least volatile—we may not even notice them to begin with—and yet they are the most tantalising and the most erotic. The base note is there to stir emotions and suggest experiences such as the mysterious East, a wood at night, and sexual encounters. Base notes smell of rare woods, damp earth, oriental spices, moss, and leather. The base notes also carry the most intimate messages of a perfume, and they even hint of baser smells such as sweat, urine and excrement. Base notes also serve an important chemical function in 'fixing' the top and middle notes by slowing their evaporation, so that together they give a more balanced fragrance over the lifetime of the perfume.

In a mixture of molecules the most volatile tend to evaporate first and the least volatile last. The wearer of the perfume wants a blend of top, middle and low notes to be released steadily and in the same proportions, otherwise their scent would change over the course of the day. The skill of the perfumers is to use the laws of chemistry to achieve this, and it can be done by carefully choosing base notes that will delay evaporation of the top notes. This is possible thanks to a curious chemical property of certain mixtures of liquids in which, given the right formulation, the volatility of each component changes so that they evaporate together and in a constant ratio.

The secret of *Chanel N° 5* is not only its chemical components but also the proportions in which these are blended. The ingredients are listed overleaf, and their relative proportions are known, but before modern methods of analysis came into use the original formula of Ernest Beaux was a trade secret.

The chemistry of smell

It is perfectly natural to want to attract other people, whether to enjoy an interesting or intimate conversation with them, or maybe to flirt.

The fragrances which make up *Chanel N° 5*

Top notes   *major:* aldehyde
                  minor: bergamot (a citrus fruit), lemon, and neroli.
                            Neroli oil comes from the bitter orange tree which grows mainly in
                            North Africa. It has a spicy, bitter-sweet odour.

Middle notes *major:* ylang-ylang
                  minor: jasmin, rose, lily of the valley, and orris.
                            Orris oil comes from the root of the iris pallida, which grows around
                            Florence in Italy. It has a warm-woody, violet-like smell.

Base notes   *major:* vetiver
                  minor: sandalwood, cedarwood, vanilla, amber, civet and musk.
                            Vetiver oil is extracted from a tropical grass and has a heavy smell
                            of earth and wood. Natural civet and musk are strong-smelling
                            secretions of animals and smell of urine and dung, but their
                            synthetic counterparts smell sweetish. The use of animal products
                            in perfumery is now insignificant.

How do we communicate these desires to those around us? Body language is one way, smell is another. Both rely very much on our 'chemistry', but using the sense of smell is a way of broadcasting this message that is *purely* chemical. The sense of smell is thought to be weak in humans compared to in other animals, and certainly smell is the least understood of the senses. It is known to be a molecular interaction which occurs in our nose. There we have sensors which can briefly trap and examine molecules that are carried on the air we breathe. If they fit the right receptors they will trigger recognition, and this will register in our brain. Clearly there are several types of receptor, and what combinations of these are activated will define the way we perceive a particular smell.

In the upper nasal tract are our olfactory nerves which consist of cilia, tiny hairlike tendrils that are too fine to be seen, but which carry an estimated five million receptor cells in the two square centimetres of the lining of the nose. The cilia are in effect bare nerve endings leading directly to the brain. Molecules are caught briefly by the fluid on the cilia, and in a fraction of a second they can move to a cavity on the surface of the receptor and be registered. How this is done is still not clear.

When they trigger a receptor, it sends its signal to the olfactory part of the brain, which in evolutionary terms is one of the oldest. From this the message passes to other parts of the brain, and in particular to the primitive limbic system which is the controlling point for mood and emotion. Part of the limbic system is the hippocampus, which is particularly important in forming long-term memory. This may explain why smells are sometimes such powerful reminders of events long ago.

Not all molecules that are volatile have a characteristic smell: the most notable exceptions are water and the gases of the air, which our noses have evolved to ignore. It is rarely of evolutionary benefit for living things to expend their energy in making or detecting molecules, unless they serve some purpose such as a sex-attractant or a warning. The smells that humans react to most strongly are those associated with food that is going bad, when invading bacteria give off the molecules **dimethyl sulfide, methyl mercaptan** and **ammonia**. Happily we rarely come across these smells from food, although we may smell them as 'bad breath' caused by bacteria in the mouth. In today's world we are much more likely to encounter fragrances specially designed to please. The addition of such chemicals to ordinary household products is now common practice, and more recently it has become part of the air conditioning of shops and hotel rooms.

When you go into a flower shop, the smell which pervades your nostrils is partly due to the blooms on sale, but also due to a commercial product that simulates how we expect such a shop to smell. High class dress shops use aromas with leathery notes to create an illusion of quality. Dr Alan Hirsch from the Smell and Taste Research Foundation in Chicago observed people shopping for sports shoes, some in a room with no smell and others in a room with a floral smell. He found that there was a marked preference for the shoes in the perfumed room.

The basic smells

By a process of trial and error, fragrance chemists have deduced a few general guidelines about what makes a molecule smell the way it does. As yet there is no exact *science* of smell, but this has not prevented chemists from adopting empirical guidelines and successfully designing hundreds of new fragrant molecules, simply by studying the structures of known examples and then copying them but with slight variations.

All hues and colours can be produced by the combination of three primary colours: red, blue and yellow in the case of paint; red, blue and green in the case of light. Colours are based on activating specific colour detectors at the back of our eyes. Taste too can be reduced to four basic flavours—sweet, sour, bitter and salt—based on four kinds of taste receptors on our tongue. Similarly there have been attempts to define a set of basic smells, and likewise we expect there to be just a few types of receptors on the cilia. The link between smell and chemical makeup is very clear because whole groups of chemically similar molecules have the same kind of smell, and this applies to groups such as alcohols, amines, aldehydes, thiols, acids and esters.

One way of classifying fragrances is according to their traditional sources. The basic categories with the largest number of fragrances notes are:

floral, consisting of oils from flowers like rose, jasmin, lilac, lily of the valley and tuberose;

green, which includes eucalyptus, pine, citrus, lavender, rosemary, camphor and basil;

animal, which consist of musk, civet, ambergris and castoreum;

spicy-wood, with extracts of oak-moss, sandalwood, myrrh, cedar, cinnamon and clove.

An early attempt to define the basic smells of perfumery had as few as five: mint, ether, floral, naphthalic (e.g. mothballs) and anise (e.g. liquorice). A later approach defined seven basic odours: camphor-like (e.g. **camphor** itself), ethereal (e.g. pears), floral (e.g. roses), minty (e.g. peppermint), musky (e.g. musk itself), pungent (e.g. vinegar) and putrid (e.g. rotten eggs).

The most sophisticated system so far devised has 14 groups, and is shown in the illustration opposite which places them in order of volatility: citrus (e.g. lemon), lavender, herbal (e.g. mint), aldehyde, green (e.g. hyacinth), fruit (e.g. peach), floral (e.g. jasmin), spice (e.g. clove), wood (e.g. sandalwood), leather (e.g. birch tar), animal (e.g. civet), musk, amber (e.g. incense) and vanillic (e.g. vanilla itself).

One day, when we know more about the receptors in our cilia, we will be able to classify all molecules according to which receptors they fit. It may also be that a particular molecule will trigger more than one type of receptor. We already know enough about receptors to realise

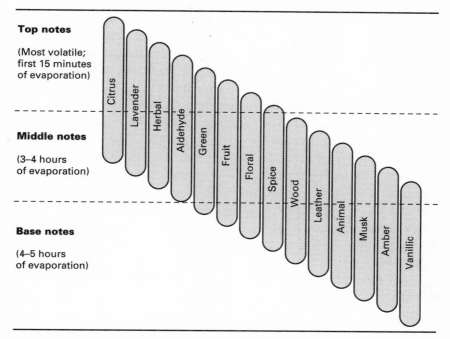

The scale of notes of a perfume. As a perfume evaporates the top notes come off first, and the base notes come off last. The perfume components shown are the 'lignes de force' of the Quest International system of fragrance classification.

that if we want to register an amber smell we need a certain shape of molecule. Chemists can study those which have the desired smell, and then make molecules that are very similar. The method sometimes works, but more often than not it fails to produce just the right fragrance combination. The difficulty with all odours is in separating the basic smell from an individual's reaction to it, which may be rooted in past experiences.

The result of research by fragrance chemists is often a new compound with its own unique aroma and nuances, and in this way they have been able to make smells that have no counterpart in nature. Such synthetic fragrances are now added to products to give them 'April freshness' or a 'summer breeze'. The skill in making these lies not so much in their smell as in their stability. Natural fragrances are complex mixtures and are often not very stable. Synthetic ones are much simpler and much more stable. They can be added to household detergents and fabric softeners because they will survive the chemical reactions of the washing process, and end up clinging to the washed clothes. This is quite a remarkable achievement because laundry products are designed to wash away unwanted materials, and they also contain highly reactive bleaches to destroy odours and stains.

Some of these artificial fragrances also find their way into fine perfumes. The problem then is to keep the price high for one kind of product, the bottle of scent, while taking advantage of its low cost to encourage its use in a large range of everyday consumer products.[3]

The price of a bottle of perfume is partly inflated by expensive packaging and advertising, but mainly determined by the degree of exclusiveness which the perfume house is trying to create for that particular fragrance. Nor should we forget that we are basically dealing with works of art, skilfully blended by a *grand nez*. The price we pay can also be justified if we feel that smelling right is important in our successful dealings with other people, be it in business or in bed. I will return to this latter topic later in this chapter when I look at aphrodisiacs.

The development of perfume

Perfumers can be seen at work on the wall paintings of Egyptian tombs of four thousand years ago, and it was clearly a well established temple industry even then. The name perfume derives from the Latin *per fumum* meaning 'through smoke', because it was common practice to send prayers to heaven by the burning of sweet-smelling incense such as frankincense and myrrh. The Law of Moses has instructions for the preparation of incense that includes myrrh and cedarwood. Incense has been burnt in places of worship for over three thousand years.

---

[3] The promoters of the respective industries have managed this with great dexterity, and there is an amazing difference in selling price between the same molecules in different packages. This is where chemistry gives way to the subtler science of economics.

Perfumes at Christmas

'Entering the house, they saw the child with Mary his mother, and bowed to the ground in homage to him; then they opened their treasures and offered him gifts: gold, frankincense, and myrrh.'

[ST. MATTHEW, CHAPTER 2, VERSE 11].

Perfume is a traditional Christmas gift, and sales of perfumes reach a peak each year in December. Frankincense and myrrh are still used by perfumers: *Timeless* by Avon and *Prélude* by Balenciaga contain frankincense; *Le Jardin* by Max Factor and *Fidji* by Guy Laroche contain myrrh. Frankincense and myrrh are resins that ooze from wounds cut into the bark of certain Middle East trees. They are not single compounds but a collection of many substances. Plants make these chemicals, and others like them such as pine rosin and galbanum, to protect their injuries against the ravages of bacteria and fungi, which would make them rot. This is partly the reason why they made good preservatives for perfumes, and explains why myrrh was also used by embalmers and in mummification.

The word frankincense comes from the Old French *franc encens*, meaning 'pure incense', but it is better known to perfumers as olibanum. It comes from boswellia trees which grow in Somalia and Saudi Arabia. Cutting the trunk of the trees causes a white juice to flow, which hardens on exposure to the air forming tear-like drops up to an inch long. These drops are harvested after 14 days. The oil that is extracted from this resin has a balsam-like, spicy note with hints of lemon.

The word myrrh comes from the Arabic *murr*, meaning bitter, and this material is a red-brown gum which exudes from thorny commiphora trees. It is still harvested in Somalia and some tears of myrrh can weight up to 250 grams (half a pound). It has a pleasant aromatic odour and was used by the ancient world in incenses, perfumes, cosmetics and medicines. It still finds use in some toothpastes, mouthwashes and tonics, and in a few perfumes.

From the fall of Rome in 410 AD until the Renaissance a thousand years later, the art of chemistry was mainly the province of the Arabs and Persians, and so was the art of perfumery. The prophet and founder of Islam, Mohammed, was a believer in the value of perfume as one of his oft-quoted sayings shows: "Three things are dear to me on Earth—women, perfume and prayer." He also made the observant remark: "Perfume is the food that nourishes my thoughts", the truth of which we are now beginning to appreciate.

The perfume industry as we know it developed in the Middle Ages and was centred at Grasse in Provence, Southern France, where the climate and soil were ideal for growing plants such as rose, jasmin, lavender, tuberose and violet, from which the fragrant essences were extracted. The traditional way of extracting oils was to place the flowers between layers of animal fat which absorbed the fragrant oil—a process known as *enfleurage*. This perfumed grease was known as a *concrete* and sometimes used as a haircream, or the fragrance oil was extracted from the fat with alcohol, to give a product known as an *absolute*.

This method was superseded by steam distillation, whereby the plant material was extracted with super-heated steam which carried off the oils. When the steam was condensed the oily layer could then be run off. A later development, introduced about 200 years ago, was solvent extraction, and the fragrance companies of Grasse specialised in this method. It involved the raw plant material being stirred in a volatile solvent, which in the nineteenth century was often **benzene**. Benzene has fallen out of use for of safety reasons, and the solvent is now more likely to be an **alkane**, and even **butane**, which is normally a gas but can be used as a solvent by working with it under pressure. After filtering the extract, the solvent is boiled off to leave behind the fragrance oil. Other solvents which have been used are **methanol**, which is a simple alcohol, and **toluene**, which is a safer alternative to benzene. Recently liquid **carbon dioxide** gas has been used as the extractant (see also Chapter 9).

The table opposite shows where some of the more common fragrances originated. Jasmin absolute is one of the most popular ones, and it has a heady, almost narcotic, smell which is ideal for middle notes of a perfume. The bitter orange tree is a particular source of fragrance oils, yielding orange from the fruit peel, neroli from the flowers and petitgrain from the leaves and twigs. Some of the oils in the table may seem out of place as components of a perfume, but pepper oil, for example, is an extract with an oriental, spicy note that is sometimes added to male fragrances.

Perfume has always been a fit offering for gods, emperors, kings, queens, and courtesans. While it was in short supply its use was

The original sources of essence oils extracted from plants

Plant material	Essence oil extracted
Blossom	rose, jasmin, tuberose
Stems and leaves	geranium, peppermint
Grasses	tarragon, sage, thyme
Roots	orris, angelica
Mosses	oak moss
Wood	sandalwood, rosewood, cedarwood
Needles	spruce, Scotch pine, cypress
Resins	myrrh, olibanum, galbanum*
Seeds	cardamom, pepper
Fruits	anise, nutmeg, coriander
Fruit peel	orange, lemon, bergamot

*Galbanum is one of the eight ingredients which God told Moses to include in incense; it comes from the resin of a Persian tree. It has such a powerful smell that if a single drop were to be dissolved in a swimming pool of water you would still be able to smell it. Its aroma is described as an earthy green pepper smell.

limited to the rich and the rulers, and this remained the case until chemistry came along. An individual plant produces very little fragrance. It takes over five tons of rose blossom to produce a kilogram of rose oil. It takes eight million jasmin blossoms to produce a kilogram of jasmin fragrance. It is not surprising that the price of each of these natural oils is several thousand dollars per kilogram.

In ages past, perfume was costly because it was rare. When Madame de Pompadour (1721–1764), the mistress of King Louis XV of France, spent the equivalent of $250 000 on perfumes, her perfumers had about 200 fragrances to choose from. Today there are over 25 000 fragrances available. Blending them into a commercial product is the art of *le grand nez*, someone who can recognise nearly 2000 different smells. It is not surprising that their skills are compared to those of great artists. A successful composition can earn a fortune.

Masculine and feminine fragrances.*

Perfume has been a part of civilized living for centuries, for both men and women, and there are differences in the fragrances which appeal to the sexes. Here is a list of the various ways into which they are divided, although some are liked by both men and women, such as chypre.

(continued)

Masculine and feminine fragrances.* *(continued)*

The classification is based on the idea that there is an overriding 'concept' which provides the theme to a perfume. The terms are generally self-explanatory, except for *chypre*, which refers to the heady scents associated with the Eastern Mediterranean, and *fougère* which means fern-like. Each 'concept' has several 'interpretations' allowing variations on the basic theme. These 'interpretations' are again self-explanatory, with *green* referring to the odour of fresh grass and leaves, *aldehydic* being the synthetic fragrances based on the chemicals called aldehydes, and *ambery* associated with the heavy and intense smell of blossoms.

Concept	Interpretation	Popular fragrances
Feminine fragrances		
Floral	Green	Chanel N° 19, Calvin Klein
	Fruity	Eau de Givenchy
	Fresh	Anaïs Anaïs, Le Jardin
	Floral	Quelque Fleurs, Fidji, Charlie, Chloé, Giorgio
	Aldehydic	Chanel N° 5, Je Reviens
	Ambery	Blue Grass, Poison, Charles of the Ritz
Oriental	Ambery	Obsession, Samsara, Shalimar
	Spicy	Coco Chanel, Opium, White Diamonds
Chypre	Fruity	Femme
	Floral–Animal	Miss Dior, Intimate, Timeless
	Floral	Ysatis, Ma Griffe
	Fresh	4711
	Green	Armani
Masculine fragrances		
Lavender	Fresh	English Lavender
	Spicy	Acqua di Selva, V by Victor
Fougère	Fresh	Drakkar Noir, Eternity for Men, Jazz
	Floral	Dunhill, Paco Rabanne pour homme
	Woody	Sandalwood, Kouros
	Ambery	Blue Stratos, Brut, Imperial Leather
Oriental	Spicy	Égoïste, Old Spice
	Ambery	Balenciaga pour homme, Relax
Chypre	Woody	Giorgio for men

	Leathery	Macassar, Aramis, Denim
	Conifery	Men
	Fresh	Aqua Velva Ice Blue
	Green	Fahrenheit
Citrus	Floral	Eau Sauvage
	Fantasy	English Leather, Onyx, Tabac Original
	Fresh	Drakkar

*Taken from the system known as the H&R Genealogies as given in *The H&R Book of Perfume* published by Glöss Verlag, Hamburg.

Modern perfumery really began this century, and looking back it is possible to see certain periods that were particularly important. These occurred as we emerged from the two World Wars, when the perfumes appeared which we now regard as landmarks. *Chypre*, launched in 1917, was eventually to give its name to a whole range of perfumes that sought to capture the heady scents of the Eastern Mediterranean. *Chypre* itself was taken off the market after a few years because it was created from components that were too limited in supply, and its manufacturers could not cope with the huge demand it generated. Its top notes were bergamot, lemon, neroli and orange; its middle notes were rose, jasmin, carnation, orris and lilac; and its base notes were oakmoss, patchouli, labdanum, styrax, civet and musk.

The table below gives other key perfumes that were typical of their generation and of which there were to be many copies. Also listed are the main fragrance for the top, middle and base notes, although of course there are several other fragrances present in smaller amounts. Once the success of *Chanel N° 5* was assured then the synthetic fragrance chemicals industry was able to contribute more and more.

Molecules with a wonderful fresh smell, a so-called 'green' note, were discovered by chemists in the 1930s; one such was called leaf alcohol, whose chemical name is **cis-3-hexenol**. This group of chemicals led to *Ma Griffe*, by Carven, introduced in 1944. Soon to follow was *Miss Dior* from Christian Dior, which combined a classic chypre with a green top note. Dior launched it to accompany his New Look fashions in the late 1940s. *L'Air du Temps* was also a success when it was introduced in 1947 by Nina Ricci, and this relied heavily on synthetic floral fragrances. It was the first 'linear' perfume, in other words it kept the same smell to the last drop. The perfume which made the biggest impression at this time was *Femme* by Rochas. Its key fragrances were peach and plum top notes, a jasmin middle note, and oakmoss and vanilla as the base notes.

The 1960s saw the introduction of fragrances designed for the 'liberated woman', and these often relied on methyl dihydrojasmonate, the man-made version of jasmin, but with a lighter, fresher odour. The 1960s also saw the launching of *Brut* by Fabergé, the first successful male fragrance to be marketed as such, rather than as an aftershave or a 'toiletry'. *Brut* contained a synthetic musk. The next decade also saw a great deal of perfume innovation when the 'lifestyle' perfumes of *Chloé* (Lagerfeld), *Charlie* (Revlon), *Opium* (St Laurent) and *Anaïs Anaïs* (Cacharel) appeared.

If plants can produce enough fragrant oils for our needs then we can grow and harvest them as a cash crop. The oils they yield are of course mixtures of molecules, and sometimes this natural blend is preferred to a blend of the pure chemical essences. It is possible to detect a slight difference between the natural blend and one made by combining synthetic ingredients. Even so the use of plant extracts has

The key notes of trend-setting perfumes

Perfume	Decade	Top note	Middle note	Base note
Quelque Fleurs	1900s	green note	rose, jasmin	sandalwood, oakmoss
Chypre	1910s	bergamot	rose, jasmin	oakmoss
Shalimar	1910s	lemon	patchouli*	opopanax
Chanel Nº 5	1920s	aldehyde	jasmin, rose	vetiver
Soir de Paris	1930s	violet	linden blossom	vetiver, styrax
Ma Griffe	1940s	gardenia	jasmin, rose	styrax, oakmoss
Femme	1940s	peach, plum	jasmin	oakmoss, vanilla
Miss Dior	1950s	aldehydes, accord	jasmin	patchouli*, amber
Fidji	1960s	galbanum, hyacinth	carnation, orris	vetiver, musk
Charlie	1970s	citrus oil	jasmin, tuberose	cedarwood, sandalwood
Chloé	1970s	coconut	tuberose	musk
Opium	1970s	aldehydes	carnation	benzoin, tolu balsam,
Anaïs Anaïs	1970s	leafy green	jasmin	cedarwood, vanilla
Poison	1980s	pimento, plum	rose, tuberose, ylang-ylang	heliotrope, vanilla

*patchouli comes from the leaves of an Indonesian shrub, and has an intense woody, sweet, balsam-like odour with spicy undertones.

declined dramatically in favour of the industrial production of aroma chemicals, and it is not difficult to see why. In Morocco, the world's largest jasmin growing area, a flower picker could pick up to 15 000 blossoms a day, working early in the morning when the oil content of the jasmin flowers is at its highest. Even so this number of blossoms is only enough to produce about 1.5 grams of jasmin absolute. World-wide demand for some fragrances like jasmin is now of the order of thousands of tonnes per year.

Making fragrant molecules

Chemists have made and recorded the properties of several million organic molecules, and many of these have a smell. However, most molecules were not made with their aroma in mind, and although reporting the smell of a compound was important to nineteenth century chemists, sadly the practice fell out of favour because it revealed nothing about the chemistry of the substance.[4] Yet for perfumers the smell is everything. The industry is still linked to natural fragrances because Nature can still come up with some inter-esting fragrance molecules, and these can provide leads to new variants that chemists can produce easily and cheaply. The two historical sources of fragrances are plants and animals.

Perfumes from plants

Once chemists identify the components of natural fragrances, they can then make modifications to them, hopefully to produce more exotic fragrances. In this way lemongrass oil can be turned to **ionone** which smells more like violets, and **eugenol** from cloves can be turned into **vanillin**, the key component of vanilla. **Limonene** from orange oil is converted to **carvone**, the major flavour component of spearmint.

Jasmin absolute contains over 200 components, and chemists have identified most of them. The main ones which play a part in the aroma, and which are named after the plant, are **cis-jasmone** and **methyl cis-jasmonate**. There are several other components which

[4] Modern health and safety guidelines advise against inhaling any untested vapour, but in any case chemists long ago ceased to report the smell of a new chemical as a way of identifying it.

contribute to the smell, such as benzyl acetate, benzyl alcohol, benzyl benzoate, cis–3–hexenyl benzoate, methyl N–acetylanthranilate, cresol, linalool, eugenol and indole. A blend of all these in the right proportions will give a good imitation of the natural product. To make an *exact* copy of jasmin absolute would in theory be possible, but it would probably cost more than the natural material because it could be difficult to make some of the molecules that are present in tiny amounts. In any case, some components have still to be identified, and it would also be rather pointless to go to the trouble of putting these in the synthetic blend if they contribute nothing to the scent.

Jasmin oil relies on a relatively few per cent of its ingredients to give it its characteristic aroma, and of the ones listed above, cis–jasmone and methyl cis–jasmonate are the main fragrance molecules. But even identifying an aroma component does not automatically mean that chemists can make them cheaper than the natural variety. However, there are ways around this problem. Cis–jasmone itself consists of a ring of five carbon atoms to which is attached a chain of another five atoms. Half way along this chain is a so-called double bond, which means that two hydrogen atoms on adjacent carbon atoms are missing. It is possible to make this molecule in the laboratory, but it requires extra steps to get the double bond in place, and to get the right molecular geometry around it. Without the double bond the compound can be made much more easily and is called dihydrojasmone—and the chemists who first made it found that it smelt the same. The double bond was not crucial to the smell.[5]

Other molecules have jasmin-type odours, and the price varies considerably. Natural jasmin extract can cost over $5000 per kilogram, chemically identical jasmone costs about $500, while dihydrojasmone costs less than $50 per kilogram. And if you want the cheapest jasmin smell of all you can get it for as little as $5 per kilogram if you use a completely synthetic aldehyde which doesn't even pretend to have any affiliation with the natural source. The scale on which these substances are manufactured reflects their price: natural jasmin production is only a few tens of tonnes per year, while over 10 000 tonnes of the lowest priced versions are made.

[5] But it can be used to show whether the jasmin oil is purely natural or adulterated with the synthetic material

The first natural aroma substance to be manufactured artificially was cinnamon in 1856, and this was made by a chemist named Luigi Chiozza. Its chief component is **cinnamic aldehyde**. In 1882 the perfumer Paul Parquet created *Fougère Royale* (royal fern) using **coumarin**, which was first made in the laboratory in 1868. Coumarin has the odour of freshly cut grass, and its addition to other fragrances produced a smell best described as fern. In 1876 vanillin, the fragrant molecule of the vanilla pod, was made in the laboratory by Professor Ferdinand Tiemann of Berlin and Dr Wilhelm Haarmann from Holzminden. They went on to become a leading manufacturer of fragrances. Cinnamon and vanillin are better known as flavours, although they are used in perfumes. Vanillin is also present in jasmin and other natural scents, and indeed it is added to the synthetic varieties as a fixative.

It was only in 1893 that the first truly synthetic fragrance molecule was patented, again by Dr Haarmann. This was ionone, which has the aroma of violets, and smells even better than the real thing. The big breakthrough in synthesis came with the discovery of the fragrant aldehydes in the early years of this century, such as **alpha-amyl-cinnamaldehyde** which has a wonderful jasmin-like fragrance, even though it is not a component of the natural material. **Hydrocinnamaldehyde** is a related chemical that gives off the heavy scent of hyacinths.

In 1885 **quinoline** compounds were first made in the laboratory, and these had the smell of leather. A natural quinoline was used in *Chypre*, and thereafter in many perfumes for both men and women: *Aramis* (by Aramis) and *Macassar* (by Rochas) are chypres for men, and they have strong animal notes although of purely chemical origin. *Diva* (by Ungaro) and *Ysatis* (by de Givenchy) are similar chypres for women.

The middle notes of perfumes were originally derived from flower oils such as rose, lily and violet. A rose fragrance can be achieved with **geraniol**, **citronellol** and **2-phenylethanol**, all cheap to make and just as nice to smell. Molecules which are derived from methyl anthanilate successfully mimic the heady, sweet notes of tropical blooms. The biggest success of the chemists has been to find molecules that smell of lilies of the valley or lilacs, both of which resisted the attempts of perfumers to extract their natural oils commercially.

But the chemists don't always have it their own way. According to Dr Horst Surburg of the German fragrance manufacturer Haarmann & Reimer, there are some scents that still cannot be made at acceptable prices compared to those extracted from plants. Equally frustrating is that natural fragrances may contain a molecule in tiny amounts that is critical to the smell. This used to pose a problem, but identifying such a component is now easy thanks to the advances in chemical analysis, even when there is as little as a hundredth of a milligram, about as much as a drop of mist.

You get what you pay for

What do we get when we buy a bottle of perfume? In terms of chemicals, the short answer is not very much. We never buy a pure fragrance chemical, or even a neat blend of several of them; they are always dissolved and diluted with alcohol as a solvent. Fragrances have to be diluted this way because the amounts we need to use are so small that even a tiny drop of undiluted perfume would be far too strong.

There is a great deal of difference in the price of a bottle of scent, depending on whether it is labelled 'perfume', 'parfum de toilette', or 'splash Cologne'. This reflects the concentration of the solution you are buying. There are three factors involved in pricing: first, the quality of the perfume essence; second, the amount of this you are getting; and third, the quality of the solvent alcohol. This list is a rough guide to perfumery terms:

	Concentration of fragrance essence	Strength of alcohol (alcohol:water)
Perfume	15%	95%:5%
Parfum de toilette	8%	90%:10%
Eau de toilette	4%	80%:20%
Eau-de-Cologne	3%	70%:30%
Splash Cologne	1%	70%:30%

Perfumes may in fact contain more than 15% of the fragrance chemicals and some have double this amount; the same is true of the other grades. What this list shows is the minimum amount of perfume that you can expect. The alcohol too varies, and here it is the extent to which it has been purified of its water content that is the primary concern. The highest grade of alcohol is not 100% as you might expect, but about 95% since there is always a small residue of water which no amount of distillation can remove. Although the 'eau' of eau de toilette and eau-de-Cologne refers to 'waters', these are still basically alcoholic solutions although they hold a significant amount of water. These are the forms in which most perfume is bought, and indeed some fragrances are only sold as eau de toilette or eau-de-Cologne.

For some people, notably Muslims, the alcohol content poses a dilemma. Mohammed, who loved perfumes, also forbade the drinking of alcohol, and although perfumes are not drunk, the alcohol is breathed in. The answer is to find another solvent, and the leading manufacturers of fragrances are now producing non-alcoholic versions.

The scents of animals

When it comes to taking raw materials for perfumes from animals, our attitudes have changed, especially when the animal concerned has to be slaughtered to provide that little bit of its body that the perfumers need. The purists may still claim to prefer natural musk to synthetic musk, but their exquisite sensitivity brings them no credit in modern eyes. The base notes of modern perfumes, the animal notes of old, must be the product of chemical synthesis. Anything else is now unacceptable.

Only four odorants come from animals: musk from the male musk deer; civet from both sexes of the civet cat;[6] castoreum from the beaver; and **ambergris** from the sperm whale.

Musk. A few plants smell vaguely of musk, such as the roots of the angelica plant, which is cultivated in Central Europe, and ambrette seeds, which come from the tropical plant *Hibiscus abelmoschus*, which are grown for this purpose. Real musk comes from the musk deer *Moschus moschiferus* which lives at altitudes of about 1500 metres in the Himalayas, China and Mongolia. This delightful little animal stands about half a metre high (20 inches) and has attractive striped markings of shades of brown on its coat. It also has a gland the size of a golf-ball at the

[6] Civet is not really a distinct class of aroma, and perfumers rank it alongside musk.

base of its anus, from which it secretes an aroma during the rutting season. A musk doe finds the smell irresistible. For centuries human hunters found it equally attractive, and even today musk deer are killed simply for this gland.[7]

In 1864 the English perfumer Eugene Rimmel, a *grand nez* of his day, reported that when hunters removed the sac from a slaughtered deer they covered their mouth and nose with linen rags to counteract the smell, which they believed was so strong that it could cause a rush of blood to the head and result in a brain hemorrhage. Even when they had gathered their harvest of musk they had difficulty in getting it back to Europe, since the fastest ships of their day, the tea clippers, refused to transport it because of its objectionable smell. At the height of the musk trade in 1900 about 1400 kg (about 3000 lbs) of musk was being collected every year, representing the slaughter of about 50 000 animals. The gland was worth three times its weight in gold by 1978, when supplies were running low, and the price peaked at $24 per gram ($675 per ounce), making a musk deer worth about $700. Real musk is still regarded as a human aphrodisiac in parts of Asia, and this is now the only market for the genuine article—but more of this topic later.

Musk has a woody, animal scent with strong hints of sexual secretions and excrement. The amount in a bottle of perfume was tiny, sometimes as little as a milligram, but that was all that was required. In some perfumes, those which relied on musk, it was of course higher. Today the amount is not limited by price, but by the effect that is desired. It would make a perfume unacceptable to many people to know it contained even a trace of natural musk, nor need it contain any, because chemists can make a perfectly acceptable artificial musk. The first such musk molecule was discovered in 1888 when the German chemist Adolf Baur made the tertiary-butyl derivative of **trinitrotoluene** (TNT) and found it smelled of musk. It became known as *Musk Baur*, and although it was a cheap alternative, it did not compare to the real thing. Even so, it led to further research of ringing the changes by attaching other atoms to TNT, and some variants were used commercially.[8]

---

[7] Today this trade is regulated by international treaty, but over 300 kg of natural musk is still collected every year, requiring the deaths of several thousand animals.

[8] Manufacture of this type of fragrance is now obsolescent on the grounds of safety.

The real musk still tantalised chemists, even when a German called Walbaum isolated the key odour component as white crystals in 1906 and named it **muscone**. It took until 1926 for another chemist, Leopold Ruzicka of Switzerland, to work out its molecular structure, and give it its correct chemical name of 3-methylcyclopentadecanone. Meanwhile other synthetic musks were appearing and Wallace Carothers, the inventor of nylon, found that by heating certain polyesters he could convert them easily to a chemical very much like muscone. His synthetic musk oil was a thousand times cheaper than the real thing, and is still used.

The main chemical components of real musk are muscone, cholesterol, **androstenone** and dehydroepiandrosterone,[9] plus traces of other more pungent molecules. Perfectly acceptable varieties of synthetic musk are now produced, and are sold under the trade names Galaxolide, Tonalid and Traseolide. Together these total about 10 000 tonnes per year.

Civet.   Unlike the poor musk deer, the civet cat did not have to die to yield up its treasure, a white fluid which it excretes into an anal sac. The civet, *Viverra civetta*, is an African cat and it can even be bred in captivity—there are civet farms in Ethiopia. Under the stress of a confined existence the cats will actually produce more civet. In its raw state civet has a very strong, slightly fecal smell, but diluted down it has a floral–musky odour and has been used in perfumes for over two thousand years. Cleopatra was reputed to be very fond of it. The chief aroma constituent is **civetone**, and this chemical is now made by the fragrance industry.

Castoreum.   This is an oily substance obtained from the anal gland of the beaver, *Castor fiber*, which lives wild in Canada, and is raised on farms in Russia. The secretion is found in the glandular sac between the hind legs. It has a warm, sensual, leather–like smell, but in the raw state is rather offensive and needs to be diluted to be appreciated, when it acquires a pleasantly sweet aroma. It is used in perfumes such as fougères,

---

[9] Androstenone and dehyroepiandrosterone are male sex hormones.

chypres and orientals, but it requires great skill to accommodate its dominating aroma.

Ambergris.   Once upon a time beachcombers could find wealth by coming across a large ball of ambergris on the sea shore. This pale grey material is produced in the intestine of sperm whales, although why it forms is not certain. Lumps the size of footballs could be found floating on the sea or washed up on beaches. It was highly prized by perfumers, and was also taken as a tonic and aphrodisiac. Ambergris floats on the sea because it is composed of terpenes and steroids, which are like hydrocarbon waxes. In the last century it was also harvested from slaughtered whales. Ambergris gives off a lingering, sweet, dry odour, reminiscent of wood, moss and seaweed.

The key component of ambergris is ambreine, which undergoes natural changes to a variety of chemical derivatives of which the one called ambergris is the key odour component. The ambergris used by modern perfumers is of synthetic origin and is sold under the trade names of Ambrox, Ambrosan and Amberlyn. These are the same chemical but produced in different ways, and from different starting materials. Most come from sclareol, which is present in clary sage. Recently, a whole new range of ambergris fragrances has been discovered, based on a very different chemical called **2-cyclohexyl-1,3-dioxan**. This bears little resemblance to the natural ambergris molecule but it has several variants which combine an amber smell with flora, woody, lily or earthy nuances.

Some day synthetic fragrances will replace all those derived from animals, and to all intents and purposes that is already the case in the developed world, thanks to the skill of chemists. Synthetic fragrances are the preferred choice, sometimes because they have unique aromas of their own, or because they are better than the naturally extracted originals and are more reliable in quality, last longer and are cheaper.

The World's largest producer of fragrance and flavour chemicals is International Flavors and Fragrances (IFF) of Union Beach, New Jersey, USA. The second largest is Givaudan-Roure, of Dübendorf, near Zurich, Switzerland, formed by a merger between Givaudan and Roure-Bertrand-Dupont. Givaudan was a company set up by the brothers Leon and Xavier Givaudan, both of whom were experimental chemists, and were based in Grasse, France. The third largest producer in the world is Quest International of Ashford, Kent, in the UK.

Human scents

Napoleon was clearly aroused by Josephine, and part of her attraction was her own personal odour. "Don't wash. I am coming home!" was the famous message he sent from his army headquarters.

According to Michael Stoddart, who is Professor of Zoology at the University of Tasmania, Australia, and author of *The Scented Ape*, there remains a biological link between the nose and the pituitary gland which controls our sex organs. Neither male nor female human sexual behaviour is triggered by specific odour chemicals—as far as we are aware—whereas many animals are known to be sexually activated by them. There is some evidence that humans respond to molecules which others are releasing, even if they cannot smell them. These molecules may even bring us attentions we do not seek.

A young woman who used a hormone treatment called cyproterone acetate to remove facial hair found herself the object of her pet rottweiler's affections. According to a report in the medical journal the *Lancet* the dog would not leave her alone and tried to mount her on several occasions, much to her embarrassment. Thinking she had an over-sexed dog she even had it castrated in an attempt to control it, but that had no effect. Little did she know that her hormone pills were acting as a powerful canine sex-attractant.

In December 1992 in Littledean, a village in Gloucestershire, England, a pig named Doris fell instantly in love with a young man delivering newspapers, and pursued him down the village street. He took refuge in a public telephone kiosk from where he telephoned the local police for help. They rode to his rescue and captured the besotted 200 pound sow. Although the young man did not realise it, he was sending Doris a message she could not ignore. He was releasing a chemical called androstenone of which she was acutely aware.

Androstenone is the sex-attractant of the wild boar. Pig farmers can buy spray cans of this substance which acts as a strong aphrodisiac for their animals. Its odour is described as musky and a bit like that of urine. The related molecule alpha-androstenol is a musky-smelling steroid found in the testes and saliva of the boar, and it will stimulate a wild sow. The truffle, an edible fungus, also produces it, which is why a sow can sniff out this delicacy even though it may be growing as much as a metre underground. There are even some sex-shops that sell it as a supposed human sex-attractant, but its efficacy is doubtful.

Androstenone is produced by glands in our armpits, and men produce much more of it than women. Humans have sebaceous glands all over their bodies, but they are concentrated particularly in the hairy parts and give off an oily liquid, especially when we sweat. This liquid contains 12 steroid-like odorants, of which androstenone is the most abundant. It seems unlikely that androstenone is a human sex-attractant because not everyone can smell it, and in a concentrated form some women are reported to find its faint aroma slightly off-putting, although there is anecdotal evidence that they may respond to it without knowing.

When one of several chairs in a doctor's waiting room was sprayed with androstenone it was observed that women were more likely to choose that chair to sit on, rather than one of the others. In another experiment in the late 1970s Tom Clark of Guy's Hospital Medical School in London sprayed unreserved seats in a theatre with androstenone and observed that these were taken in preference to unsprayed seats. Androstenone can be detected at one part per billion, and women can be ten times more sensitive to picking up its smell than men. Ovulating women can even detect the smell at one part per trillion. One theory is that humans may react to this molecule because it is the smell we associate with close human bonding, and that we learn this association when we feed at our mother's breast, because this is the area of a woman's body which also gives off androstenone.

Nowhere is the importance of chemistry more apparent than in the sexual communication between living things. For insects, there are pheromone molecules which convey messages about food trails or act as alarm signals, but the best known are the sex pheromones, the molecules that convey the presence of a female to a male and bring him winging to her. This behaviour is built in to the insect's genes. When an animal is ready to reproduce then it needs to broadcast this information, and what better way than to release a few molecules into the environment, molecules that will be picked up by the receptors of potential mates and which at the same time switch them on sexually. Insects can be caught by traps baited with sex-attractants.

Few of the smells that humans produce are attractive. Some of the worst-smelling are the sulfur compounds which are the result of bacterial action on our body's waste products. Our sebaceous glands start working at puberty and discharge an odourless oil which bacteria can work on and convert to fatty acids which smell unpleasant. Bacteria can do the same to butter, hence the similarity between the smells of stale sweat and rancid butter.

The strongest smelling molecules from human skin are excreted by the apocrine glands, which are located in the armpits, the scalp, and around the sex organs and anus. The chemical responsible for the offensive odour of human armpits was discovered in 1991 by Dr George Preti of the Monell Chemical Senses Center in the USA. It is a substance called **3-methyl-2-hexenoic acid** (MHA) and was extracted from the sweat of male volunteers who wore absorbent pads under their arms for 24 hours. Dr Preti actually identified almost 40 molecules, and then made samples of each of these in the laboratory. As soon as he made MHA the smell of armpits was obvious. The chemical is formed by bacteria whose enzymes attack a protein from the apocrine glands, of which men have many more than women. The protein contains chemically bound derivatives of MHA which the enzymes release. Most deodorants work by blocking the glands which release the protein, but a new deodorant has been patented which works in a different way: by attacking the bacteria and stopping the release of MHA.

Having eliminated unpleasant body odours by daily showering and using deodorants, we no longer need to seek refuge in perfume to mask smells as did previous generations. Now we can enjoy perfumes for their own sake. Nevertheless a good perfume is more than just a nice smell, and our knowledge of what the deeper notes of perfume are saying may incline us to believe that there is a strong link between perfumes and sex. They may even be regarded as mild aphrodisiacs.

Aphrodisiacs

Perfumes may not be true aphrodisiacs, but they come very close, and they have always been the stock in trade of the professional purveyors of sex.[10] Throughout history prostitutes have been highly perfumed, and some were even more subtle. Parisian courtesans of the last century were reputed to wear a bag of musk between their breasts. Aphrodisiacs are generally regarded as something to be taken by mouth rather than merely sniffed. It may be true that chemicals may make us smell much more alluring that we would otherwise be, but it is not true that they can improve our physical performance. Until we realise this then other species will be driven to extinction—and no amount of chemistry can prevent it.

When an Indian bull rhino mates with a female, the sex act lasts an hour and during this time he can ejaculate a dozen times or more. Little wonder then that the rhino is regarded as a potent sex symbol, and Chinese doctors prescribe rhino horn as a natural medicine for impotence as well as such ailments as fever, arthritis and lumbago. A couple of grams of rhino horn shavings cost about $30, making a complete horn worth several thousand dollars. Indian rhino horn commands a premium price many times that of the African rhino.

Rhino horn is sold in many parts of the East as an aphrodisiac even though it has been shown scientifically to be useless for this purpose. Drinking rhino horn tea, the traditional way of taking this love potion, gives you the same boost as making tea from your own nail parings. Rhino horn is keratin, the same chemical that pigs grow as trotters, cows grow as hoofs, and humans grow as fingernails. International groups such as the World Wide Fund for Nature are now attempting to stamp out the sale of rhino horn because in Africa there are fewer than 5000 rhinos left, and in India the numbers are down to 2000.

Love potions have been sought since the dawn of time, and there are scores of natural products that have the reputation for increasing our sexual appetite. Alcohol and marijuana are regarded as mild aphrodisiacs, but that is because they make us less inhibited. Some foods have this reputation as aphrodisiacs because of a resemblance to human sex organs: bananas, asparagus and the stinkhorn mushroom

[10] Aphrodisiacs are named after the Greek Goddess of love, Aphrodite.

have the shape of an erect penis; mussels, figs and oysters look like female genitals.

Casanova, the 18th century lover, certainly thought food could be an aphrodisiac—he ate 50 oysters in an evening. There may even be a scientific explanation for his predilection for these molluscs. A key trace element in our diet is zinc, which is essential for growth and is a constituent of many key enzymes of the body, especially those which synthesise protein. Zinc is very important for the maintenance of the sex glands, and in the hormones that govern our sex-drive. Semen has a high level of zinc, and this continually needs to be replenished. One food rich in zinc is oysters, so Casanova was doing himself no harm in eating lots of them: a pound of oysters provided him with 120 mg of zinc.

There are scores of other foods that are reputed to have aphrodisiac properties: among them are chocolate, oats, sunflower seeds, nuts, avocados, carrots, celery, mangos and fish. Garlic is claimed to be an aphrodisiac in France, but in Greece it is thought of as a passion-killer. None of these foods contains any chemical that has been proved to be effective, but some, like chocolate, contain **phenylethylamine** which is thought to be linked to sexual arousal in the brain, and it has similarities to natural chemicals that act as stimulants, such as amphetamines.

Some of these foods provide us with essential nutrients: for example carrots are rich in vitamin A, and this is important for converting cholesterol into the male hormone testosterone. If he does not get enough vitamin A a man becomes sterile, but he is rarely in danger of not taking in enough of this vitamin in a normal diet. Indeed he is more likely to take in too much and provoke a toxic response. The aphrodisiac properties of food are imaginary, and the Sexual Research Institute at Hamburg University failed to find any 'active' ingredient in over 200 supposed aphrodisiacs. If popular aphrodisiacs work it is really because they act on our imagination.

Very few true sex stimulants are known to science. There are some spices that seem to have an effect, such as chili peppers, but this probably is because they contain irritating chemicals that cause a tingling sensation in our bowels or bladder in the same way that they burn our mouth and tongue when we eat them. The chemicals **yohimbine** and cantharides cause an erection in men as the outward

sign that they are having an effect on the body. Whether this means
they are aphrodisiacs is a debatable point, since they do not stimulate
the desire for sex, only the ability to engage in sexual activity.
Yohimbine is a crystalline compound that comes from the bark of
the yohimbé tree (*Corynanthe yohimbé*) which grows in central Africa.
People there have used it for centuries to stimulate their sexual
powers. It dilates the blood vessels of the sex organs, leading to an
erection in men and increased sensitivity in the genitals of women.
Scientific research on rats and humans has proved that the effect is
real. As little as 10 mg of yohimbine is enough, and too much is
dangerous. A man who injected himself with 1800 mg collapsed in
a coma and only just survived.[11]

Spanish fly, the ground-up powder from the brilliant green beetle
Lytta vesicatoria, is the source of cantharides, but there are risks because
this chemical can also kill. In one case in the 1940s a young man
working with a London firm of medical suppliers laced some coconut
ice with it and gave it to his girlfriend. After two days of agony she
died, her intestines eaten away by this corrosive irritant. A small dose
of Spanish fly does produce a strong erection in men, the effect being
due to its irritant effect on the urethra, the tube that runs from the
bladder to the tip of the penis.

Arsenic is a deadly poison, but in small doses it gained a reputation
as a tonic and was especially popular with Victorian businessmen.
Even Charles Dickens resorted to it. It probably got the reputation
as an aphrodisiac because it was used in medicine to treat tuberculosis.
In the final stages of this wasting disease male patients become notor-
iously sexually aroused, and this became associated with Dr Fowler's
Fever Cure, which many of them were given as treatment. The
famous fever cure was a solution of potassium arsenate, a few drops
of which were taken in a glass of water. In many parts of Britain it
became common practice for pharmacists to dispense doses of the
tonic over the counter. It may have had some effect because a small
dose of arsenic will boost the body's metabolism, making people feel
more alive, and so more sexually alert. The arsenic factor may also
explain why certain foods with high arsenic content also gained the
reputation of being aphrodisiacs, such as plaice and shell fish.

[11] The fatal dose is 3000 mg.

Aromatherapy

Perfumes may not be aphrodisiacs but they can have strong psychological benefits, and this is the philosophy behind aromatherapy. Because so few of the fragrant molecules are actually trapped by the invisible hair-like cilia in our nostrils, it is unlikely that they can have an effect on our bodily processes, but they may have an effect on our emotions, in just the same way that music, colour and the spoken word can affect us deeply. Extracts of the sebaceous gland have been tested at stress clinics and found to have a calming effect, and the reason appears to be that we associate them with the time we felt most secure in our lives—when we were being suckled as a baby.

Research at the Toho University School of Medicine in Tokyo indicates that smell can have a profoundly moving effect. Electrodes attached to a person's scalp were used to monitor changes in brain waves as the person was exposed to different aromas. Jasmin, for example, acted as a stimulant, just as aromatherapists claim, while lavender calmed brain activity down: this fragrance is recommended as a sedative. The researchers also found that the aromatherapists correctly predicted the stimulant or sedative readings of various smells—except that of rose oil. Aromatherapists speak of rose as a sedative, whereas according to the brain-wave measurements it is a stimulant. Green notes such as mint and rosemary improve concentration; citrus notes make us feel more alert; lavender relaxes us; rose, jasmin and ylang-ylang make us think of love; while musk and sandalwood are decidedly warm and sensual.

The Fragrances Foundation in New York, headed by Annette Green, is researching and promoting smell and drawing up lists of behavioural fragrances. Mood perfumes are set to become part of everyday living. Peppermint is said to increase production and reduce accidents in factories, while vanilla comforts patients in hospital, and floral scents make people spend more time, and money, in shopping malls.

We have every reason to believe that the golden age of fragrances is only just beginning. Research in chemistry will continue to turn up yet more wonderful smells, and perhaps psychologists and medics will be able to use them as the most gentle of mood-modifying chemicals. The philosophy of perfumers and aromatherapists may one day turn into a science of chemistry and mind. Even the major perfume houses are starting to market products with emphasis on mood. For example *Wings* (Giorgio Beverly Hills) claims to make you feel 'uplifted', *Champagne* (YSL) to make you feel happy, and *Tuscany Per Donna* (Aramis) to make you feel seductive. They might be right.

CHAPTER TWO

Sweetness and Light

Sugar; honey; sugar as a chemical resource; bulk sweeteners; artificial sweeteners; what makes a molecule taste sweet?

IN the previous chapter we looked at fragrance chemicals, which we are happy to use to improve the quality of our lives, and in the knowledge that they no longer involve the needless slaughter of wild animals. Advances in chemistry have brought this about, but advances in other areas of chemistry have not led to similar acclaim. One such area is the production of chemicals that are put into foods. Indeed the very idea of *eating* chemicals seems wrong, and yet we do it every day.

Of course I could point out that every mouthful of food we swallow is merely a collection of molecules, but this can be countered by pointing out that food molecules come from other living things and are part of a natural food chain. When people speak of something as 'chemical' they often mean it as a criticism of something unnatural, originating as it does in a chemical laboratory. Sometimes this approval of things natural and condemnation of things chemical is turned on its head, and this is what has happened with the sweet-tasting components of our food. Sugar, which is entirely natural, is seen as a threat, while artificial sweeteners are booming, although they too are considered a health hazard by some people. When we look a little closer, however, we find that there is little to worry about in either case. The dangers of sugar are much exaggerated, and it has much to offer the world.

We like sweet-tasting foods, and we have specific sweetness detectors on our tongue that register this pleasant sensation. So what exactly is sweetness? And why do relatively few natural substances have this property? The answers to these questions are to be found in chemistry. If we take the molecule's eye view of sweetness we can perhaps understand better the controversies that surround both white sugar and artificial sweeteners.

Food chemists divide sweeteners into two kinds: bulk sweeteners and intense sweeteners. Sugar is the best known bulk sweetener; it is sweet but we require a lot of it to achieve the degree of sweetness we want. Bulk sweeteners may also give us something we don't want, and that is excess Calories.[1] Intense sweeteners avoid this problem, but some people believe they may be dangerous in a different way because they threaten our health. To put it rather simply: eat too much sugar and you end up overweight, with tooth decay and possibly heart disease; eat too much **cyclamate**[2] or **saccharin** and you might end up with cancer. Either way it seems we have to pay a heavy price for indulging a craving for sweet foods. Or do we?

Sugar

The sugarcane plant is native to Polynesia. It was taken from there first to China, and then to India, where it was refined into sugar in about 700 BC. Slowly the cultivation of sugarcane spread westward through Persia to Egypt, and was to be found around the warmer parts of the Mediterranean by the early Middle Ages, around 800 AD. Muslims were particularly active in growing it and appreciated the sweet-tasting sherberts made from it. These they drank in place of alcohol which had been forbidden by The Prophet.

---

[1] A calorie is the energy needed to raise the temperature of a gram of water by one degree Celsius. It takes about 50 000 calories to boil a pint of water. Food chemists and dieticians talk in terms of the kilocalorie, which is 1000 ordinary calories, and they would say that it takes about 50 kilocalories to boil a pint of water. Sometimes kilocalorie is written as Calorie with a capital C, and this is the common unit for the energy value of food.

[2] Where the name of a substance is given in **bold** type, it means that there is more information about its chemistry in the Appendix.

The first sugar reached England in 1319 AD, but it was an expensive novelty to begin with and used only in medicines. It had to be imported and cost ten times as much as locally produced honey. Nevertheless there was a demand for it for special occasions, and in Shakespeare's *The Winter's Tale*, which was perfomed in 1611, the Clown asks: 'What am I to buy for our sheep-shearing feast? Three pound of sugar, five pound of currants; rice—what will this sister of mine do with rice?...' Sugar continued to be expensive for people living in northern Europe throughout the Middle Ages and right up to the seventeenth century, when cheaper imports from the Caribbean began to arrive.

Sugarcane is a jointed reed standing 2–3 metres tall (6–9 feet), and a plant will grow for up to ten years in deep, rich and well-watered soil. It will not survive frost. Sugarcane shoots were transported by Columbus to the New World on his second voyage in 1494, and there it thrived. The climate of the West Indies was ideal, but even so it took two hundred years before it was being exported back to the Old World.

Cutting sugarcane is hard work and the notorious slave-trade developed to provide planters with cheap labour to harvest this twice-a-year crop and to work the vats in the sugar mills, where temperatures regularly reached 50 Celsius (over 120 Fahrenheit). Sugar is easy to extract from the crushed cane because it dissolves readily in hot water. On cooling the solution the first crystals to appear are pure and white, while subsequent crops of crystals become progressively darker. This brown sugar is referred to by a variety of names, the most common being demerara. Finally a black tarry liquid, molasses, remains which is still about a third sugar: the remaining two thirds are other carbohydrates, various acids, salts and amino acids. Molasses are fermented today to produce chemicals such as **ethanol**, **acetone**, **glycerol** and **citric acid**. It can also be turned into rum or used to grow baker's yeast.

The impetus for change in the sugarcane industry came as a result of the Napoleonic Wars and the blockade of France by the British Navy, which cut off supplies of sugar from the French colony of Dominica (Haiti). Napoleon set his scientists to seek alternative sources and they took up the researches of Andrew Marggraf, a chemist at the Berlin Academy. Marggraf had discovered that sugar could be extracted from root crops such as parsnips, but he was particularly impressed by the amount that came from the sea beet, a cousin of the beetroot. Selective breeding of sea beet to maximise its sugar content began about 1800, and within ten years a high-yielding variety was developed and the beet industry was born. The French Government encouraged farmers to grow sugar beet so that the country would never again be dependent on imported sugar from the West Indies. Sugar-beet was ideal for the cooler climes of Europe, and today it is a major crop around the world. Indeed, it outstripped sugar-cane as the main source of sugar over a hundred years ago.

White sugar is just one of a group of sugars. To put it more exactly: **sucrose** (which is another name for sugar) is a simple carbohydrate. Or, to put it in chemical terms: saccharose (which is the technical name for sugar) is a di-saccharide. There are many saccharides.[3] This group of compounds ranges from the simplest so-called mono-saccharides, such as **glucose**, up to the large poly-saccharides, such as starch and **cellulose**, in which there are long chains of hundreds of saccharide units strung together.

Glucose is the most abundant monosaccharide and is the first product of photosynthesis in plants. It is present in many fruits and vegetables, and is the main source of energy for both plants and animals. Other monosaccharides are **fructose** which is also found in fruits, and **galactose** which is produced by animals.

When two of these simple sugars join together we get a disaccharide, and sugar itself is the combination of glucose and fructose. Maltose, the saccharide of malted barley, is also two units, both of them glucose. Lactose, the saccharide of milk, is glucose plus galactose. Other common saccharides are fucose, which is a component of mucous, and **xylose**, which is found in plant gums. The most abundant carbohydrates in plants are the polysaccharides starch and cellulose. Both consist of many glucose units, but these are connected together slightly differently, with the result that we can digest starch by breaking it down into glucose, but we can not digest cellulose. Consequently starch is a rich source of food energy, while cellulose

---

[3] The technical term saccharides derives from a Medieval Latin term which in turn came from the Greek *sakkharon*, meaning sweet.

passes through us unchanged, and is better known as dietary fibre (see Chapter 4).

Glucose is also known as 'blood sugar' because it accounts for 0.1% of our blood. Our body tries to keep the level of this constant. The glucose derived from our food, by the action of enzymes in our stomach, can pass directly into our blood stream and be taken to where its energy is needed. As it is used, it breaks down into simpler molecules and eventually ends up as carbon dioxide and water,[4] a process which releases a great deal of energy for our body, either as heat or muscular action. When the level of glucose in our blood rises after a meal we turn the excess into another polysaccharide called glycogen, which we store in our muscles and liver. There it waits until the glucose level in our blood falls again, and it can then be turned back into glucose and used as required.

The average adult has about 350 grams (three quarters of a pound) of glycogen stored in their tissue at any one time—and this has energy to supply us with about 1400 Calories, enough to keep us going for a day. If we eat more food than we need, and the glycogen reserve is already full, then our body starts to turn glucose into fat, which it can store more economically and of which we have a bigger reserve—enough to keep us alive for a month. We will come back to the problem of running down this energy store in Chapter 4.

Sugar is also easily digested, and provides half its weight as glucose. The other half, fructose, is an equally good source of energy, although it is absorbed more slowly. Fructose actually tastes sweeter than sugar, but it was the fructose part of sugar that some suspected of causing ill health.

Earlier generations viewed sugar as a wholesome food, valued for its purity, and much used in the home for baking cakes and apple pies, or for making toffee and jam. Sugar is nice, but there has always been a price to pay. Dietitians, dentists and doctors warned their patients about eating too much sugar. Dietitians told their obese patients to give up sugar as part of a weight-reducing diet. Being overweight can have serious effects on our health, leading to less activity, high blood pressure and heart disease. Dentists knew that sugar caused tooth

---

[4] The overall reaction is chemically the same as burning it in air and thereby reacting it with oxygen.

decay, which it does by penetrating the film of plaque which covers our teeth. Plaque bacteria, and in particular *Streptococcus mutans*, break down the sugar and form acids like lactic acid which then attacks the **calcium phosphate** that makes up the enamel surface of the tooth. A cavity forms, which might eventually etch its way to the pulp part of the tooth, causing pain and allowing infection to spread deep into the tooth and even down to the root, where it results in an abscess.

Doctors once warned patients with diabetes to avoid sugar at all costs since they lacked the insulin necessary to utilise glucose and release its energy. Before insulin was available on prescription, a person with diabetes was advised to eat a high-fat, low-carbohydrate diet, since their body could digest fat but not carbohydrate.[5] A diet comprising at least two-thirds fat was the target to aim for. Many ate special diabetic food if they wanted a sweet-tasting treat, and these were made with bulk sweeteners like **sorbitol**, a simple sugar made from glucose. However, a person with diabetes who can produce some insulin, but not enough, does not have to avoid sugar assiduously; indeed this would be almost impossible since a lot of ordinary foods contain a little sugar. The modern approach to diabetes, which for many people can be managed by controlling the diet, is not to rely on sugar substitutes: such people are advised to watch what they eat, and of course to avoid eating foods made with lots of sugar. Dietary advice for diabetes is now high-fibre and low-fat, and to eat some bread and potatoes, food that earlier sufferers were once warned against. Of course not all diabetes can be managed by diet, and many diabetics have to rely on injections of insulin.

A whole new range of diseases for which sugar was to blame began to emerge in the 1960s. The public became aware of these with the publication in 1972 of the catchily titled book *Pure, White and Deadly*, written by Professor John Yudkin, a nutritionist at Queen Elizabeth College, London.[6] Yudkin's book blamed sugar for causing diabetes and ulcers and promoting heart disease. He claimed that sugar was not

---

[5] Strictly speaking people with diabetes can digest carbohydrate, but to stimulate cells to take up glucose and use it requires insulin.

[6] The book *Dietetics, Coronary Thrombosis and the Saccharine Disease* by T.L. Cleave, G.D. Campbell and N.S. Painter appeared in 1969, and was the first to warn of the dangers of sugar.

like other carbohydrates: 'recent research shows that it also has unique effects in the body, different from those of other carbohydrates.' Moreover, and more startling, was Yudkin's assertion that 'if only a small fraction of what is already known about the effects of sugar were to be revealed in relation to any other material used as a food additive, that material would promptly be banned.' He could not explain exactly how sugar caused these various diseases, but he speculated that it might do it in several ways, such as by swelling organs in the body, or by causing the over-production of hormones like insulin, or maybe by boosting the population of microbes in our intestines.

He wrote off sugar as merely 'empty calories', the implication being that it provided nothing but energy, and indeed sugar has none of the other components we need for a balanced diet, such as protein, vitamins and minerals. We need these to repair our body and keep it functioning effectively. The term 'empty calories' could equally be applied to all carbohydrates, including glucose, fructose, lactose and starch. Yet carbohydrates are not used by the body solely to generate energy: some are needed to build essential structures within the body, and some end up attached to proteins or to cell membranes, where they play an important role in the process whereby one living cell recognises another. To label sugar and its components as merely empty calories was rather unfair. Even so, the amount of sugar in the average person's diet was far in excess of the body's need for cellular building blocks, and of course most sugar is used as energy or turned into fat. The problems of sugar were really the problems of being overweight, and this was often to blame for most of the accusations being made against sugar.[7]

As the years went by the list of ailments which appeared to be linked to sugar grew to include not only heart disease, but also appendicitis, gallstones and hernias, until by 1983 the Royal College of Physicians in London felt constrained to recommend that the national intake of sugar be reduced by half. By this time sugar was even accused of causing breast cancer and varicose veins. In 1986 a more rigorous study of all the charges against sugar was

[7] Later the blame was to pass to fats in the diet, and this aspect we will consider in Chapter 5.

carried out by the US Food and Drugs Administration (FDA). This body reported that it could find no evidence for singling out sugar as the primary cause of ill health. Nor was this conclusion surprising, because the glucose and fructose of sugar are no different to that of other foods.

However, the years of bad publicity had had their effect. People stopped putting sugar in coffee and tea, yet surprisingly the amount of sugar consumed did not change significantly. People ate it in other ways, such as in biscuits, snacks and confectionery. The average person in the West still eats around 100 grams (4 ounces) of refined sugar a day; in some countries, such as Australia, sugar consumption is a third higher than the average, while in others, like Japan, it is a third lower.

So how should we think of sugar? Should we do all we can to avoid it? In fact it would be almost impossible to cut out sugar completely from our diet, because that would mean eating no fruit or vegetables. In all respects, both chemically and nutritionally, sugar molecules are the same whether they come from sun-dried raisins or from a sugar refinery. Apples have 2% sucrose, 2% glucose and 5% fructose; oranges have 5%, 2% and 2% respectively; bananas have 7%, 2% and 1%; and grapes have no sucrose but 8% each of glucose and fructose. The total amount of these sugars in grapes rises to 64% when they have been dried to make raisins or currants. Dried dates are 73% sugars. Half of the sugar we eat is refined sugar; the other half is naturally present in foods. Together they provide us with about 20% of our daily energy requirement.

A lot of refined sugar is eaten in cakes, biscuits, fruit pies and confectionery, which are obviously sweet to the taste, but a lot is eaten in other foods which we do not think of as predominantly sweet, such as savoury snacks, bread, pickles, sauces and convenience foods. Some people may have stopped adding sugar to tea and coffee, but they may take more than they suspect in fruit-flavoured drinks, mixers and colas, where it is used as a balance to the acidity.

As a natural resource, sugar is pure, white and *heavenly*. Refined sugar has some wonderful advantages: its chemical purity means we do not need to worry about contaminants; it is easily digested and provides energy in abundance; and it is stable and acts as its own preservative. Sugar will last for years if it is kept cool and dry. Any country whose agriculture fluctuates between feast and famine would be well advised to store sugar as insurance against the lean years.

The simple equation linking weight and energy explains why this is so. Bare survival requires us to take in 1000 Calories of energy every day, which we can get from about 280 grams (10 ounces) of sugar. If this were the only source of energy in our diet, a normal person would need 102 kilograms (228 pounds) a year to stay alive. A million tons of sugar would provide enough *energy* to keep about ten million people alive for a year. Fats and oils, which weight-for-weight provide over twice the energy, might seem better foods to store, but they are susceptible to attack by oxygen in the air and go rancid, although this can be delayed by adding antioxidants.

Of course more than sugar is needed to sustain life, and a famine diet still has to include some essential **amino acids** from proteins, plus vitamins and minerals, without which the body cannot make new cells nor keep existing cells functioning. A diet to get you through a famine must include at least some of these, and this is why skimmed milk powder, which is rich in protein and minerals, is also considered a good stand-by food—although not for everyone. Adults in some ethnic groups cannot digest the lactose in milk. Dried egg would be an alternative source of easily stored protein.

Meanwhile in the overfed economies of the West, people fight the twin scourges of sugar and artificial sweeteners. To help them in their struggles, products in supermarkets are labelled 'no added sugar' or 'contains no artificial sweetener', although the former might well contain natural sugar and the latter almost certainly contain the refined variety. Some people even try to do without sweet-tasting foods altogether, in the mistaken belief that if something tastes sweet it must be dangerous to health. The only exception they make is honey, which of course is beyond criticism, and is often labelled 'pure'[8] and 'natural'.

---

[8] The phrase 'pure honey' refers not to the chemical composition, but to the source of the honey, and means that it is not a blend of honey but is from a particular apiary or local area. It does not mean the same as the 'pure sugar', which does refer to its chemical composition.

Honey

Many who oppose sugar advocate honey as a healthier alternative, but a closer look at this food shows that they are deluding themselves. Honey is basically just a collection of sugars.

Honey is as old as history. There is a reference to honey bees in ancient Egypt 5000 years ago, and honey is mentioned several times in the Old Testament of the Bible, where its sweetness was proverbial. Today over a million tonnes of honey are produced worldwide, the main exporting countries being Russia, China, the USA and Mexico. A colony of bees can produce up to 70 kg of honey (about 150 lbs) in a season. Honey is gathered by billions of bees plundering the nectar from trillions of flowers. Nectar is mainly a sugar solution, and when this is taken back to the hive it breaks down into glucose and fructose.

Honey is a collection of saccharide chemicals and water. Glucose and fructose together make up about 70%, sucrose and maltose about 10%, and water accounts for the remaining 20%. There are tiny amounts of many other components, and these provide the characteristic aroma and flavour. Over twenty other saccharides have been identified in honey. There is also gluconic acid, and minute traces of proteins and minerals, but there is so little of these that they do not make honey a good dietary source for anything but carbohydrates. If sugar is to be derided as 'pure, white and deadly' by those who oppose it, then logically they should refer to honey as 'impure, golden and deadly'.

Honey is regarded by many as a health food, and spoken of as if it had almost magical qualities. Unlike pure sugar, honey has a comforting smell and a taste all its own, and these come from 120 aroma chemicals that are present. Food chemists have identified most of them, but about 40 still remain to be classified. Some of these chemicals come from the flowers from which the bees have gathered the nectar, and it is possible to identify lavender honey or heather honey if the bee hives have been next to a large area of these plants. One flavour molecule dominates and gives honey its characteristic smell, and that is ethyl phenylacetate.

In effect honey is mainly **invert sugar**—and so is golden syrup. The resemblance is more that just superficial. Invert sugar is the name given to ordinary sugar when it has reacted chemically with water and split into glucose and fructose—which is what happens to nectar in the hive. Add a drop of the flavouring ethyl phenylacetate to invert sugar, and you would be hard pressed to tell it from honey.

All the sugars in honey can be broken down to alcohol by fermenting them with yeast, and this makes mead, which was a drink of the rich for hundreds of years, especially in northern climes where grapes could not be grown. Mead continued to be made on a large scale in Russia even in the last century, and it is still produced as a novelty drink in England today.

Sugar as a chemical resource

Sugar makes a superb agricultural crop, quite apart from its use as a human food. Sugar-cane and sugar-beet can be grown as sources of renewable energy, and sugar can be used as a resource for the chemicals industry. The world produces about 120 million tonnes of sugar per year, and could produce much more in both tropical and temperate regions. Brazil was the first country to grow sugar-cane on a large scale to make fuel for transport, in this case to make alcohol which today powers millions of cars. We will return to this aspect of sugar in the next chapter.

Sugar can also be used as a chemical in other ways, not by breaking it down to simpler compounds like alcohol, but by using the sucrose molecule itself and adding other groups of atoms to it. The company Procter & Gamble turns sugar into a fat substitute which they call Olestra. Other chemical companies are using sugar to make detergents, pesticides and even an anti-cancer agent. Attaching eight **acetate** groups to sucrose creates one of the most bitter compounds known, sucrose octa-acetate, and this can be used to give a liquid a nasty taste so that people are warned against drinking it.

Except where they are destined for human consumption, few of these sugar-based products have achieved commercial success because they compete with oil-based chemicals, which are much cheaper. But if there ever comes a time when fossil fuels run out, we could turn to carbohydrates like sugar to make the hydrocarbons we now get from oil. Hydrocarbons are the starting point for most of the industrial chemicals used to make plastics, resins, fibres, solvents, dyes, paints, drugs and adhesives.

We could get hydrocarbon petrochemicals from sugar by two routes. One way would be to ferment the sugar to alcohol, and then take this one step further by heating to a high temperature to convert the alcohol to the more useful gas ethylene. This is the starting point for materials such as plastics and resins (see Chapter 6). Another way would be to ferment sugar with the micro-organism *Bacillus acidi levolactiti* which produces lactic acid. When this chemical is heated in water under very high pressure it forms **acrylic acid** from which acrylates can be made. These are used in paints, metal coatings, fibres and polishes.

Other micro-organisms can work on sugar to yield such products as **acetone**, **butanol** and **2,3-butanediol**, all of which have a role in the chemical industry. Sucrose is already used to manufacture other consumables apart from alcohol, and the amino acids, L-glutamic acid and L-lysine come from the sucrose of molasses. L-Glutamic acid is best known as its sodium salt, the flavour enhancer monosodium glutamate (MSG). Other industrial processes now in use convert sugar to glucose, and ferment that to make the vitamins B_2 and C. Glucose can also be fermented by the moulds *Martierella* or *Rhizopus* to make γ-**linolenic acid**, which commands a high price as evening primrose oil.

The most environmentally friendly product made from sugar is the biodegradable plastic polyhydroxybutyrate (PHB). The chemical company Zeneca produces this from sugar and markets it under the name of Biopol, which has won environmental awards in the USA and Germany. The German hair care company Wella sells its Sanara brand of shampoo in bottles made of Biopol, and they reported increases sales when these were introduced. Biopol is biodegradable because PHB makes an ideal food for other microorganisms, which dispose of the plastic by digesting it. In so doing they return it to the air as carbon dioxide.

PHB can be modified by adding another hydroxy-acid such as 3-hydroxypentanoic acid, and in this way the texture of the polymer can be tailored to make it suitable for plastic envelopes or carrier bags. However, PHB is expensive, and a container made of Biopol is several times more expensive than the same container made of PVC or polythene. But its cost can be justified for some uses, and 5000 tonnes a year are manufactured. PHB will not solve the problem of plastic litter because it is too expensive to be used for throw-away packaging. Its main use is likely to be for surgical sutures and supports within the body where it can slowly dissolve during healing, or for medical and personal hygiene products that we flush away to the sewers.

Bulk sweeteners

Bulk sweeteners, as their name implies, need to be used in relatively large amounts to give you the desired level of sweetness. Glucose, **sorbitol** and **xylitol** are bulk sweeteners, and sugar is *the* bulk sweetener—others are generally less sweet. Sweetness is measured by making a 10% solution of the compound in water and then asking a panel of people to taste it. It is then diluted, tasted again, diluted, tasted again, and so on until they can no longer taste it. This way you can compare sweetness, but it is still rather subjective. The values in the table below are only a rough guide, and sugar is used as the standard against which the others are judged. The values fall in a narrow range, and you should bear this in mind when we consider artificial sweeteners later in this chapter, some of which are over a thousand times sweeter than sugar.

Golden syrup can be made from sugar either by the action of enzymes or acid, and it is only one type of liquid bulk sweetener. Others can be made by doing the same to starch from corn, potatoes or wheat, and the glucose syrups they produce are used when a liquid

Sweetness factors compared to the sweetness of sugar	
Fructose	120%
Xylitol	100%
Erythritol	75%
Glucose	70%
Galactose	60%
Maltose	50%
Sorbitol	50%
Lactose	40%

sweetener is required for ice cream, chewing gum, chewy sweets or candy. We could even turn **cellulose** (fibre) into a sweet-tasting nutritious syrup, but the process is more difficult and not economical at the moment. If there were ever a severe worldwide famine, chemists could increase the food supply by turning sawdust and paper into life-saving 'empty calories' of glucose.

It is possible to turn sugar into other sweet-tasting saccharides apart from glucose and fructose. There are enzymes that will take sucrose and rearrange the molecule to a form known as isomaltose, which is also sweet but nowhere near as sweet as sugar itself. The saccharide which comes nearest to meeting the requirements of a bulk sweetener that is as sweet as sugar is xylitol.

Xylitol

Xylitol was first made over a century ago in Germany by Emil Fischer and his associates. It has a sweetness rating about the same as sugar itself, and it has as many calories, but it cannot be digested by the bacteria that cause tooth decay. Xylitol penetrates the plaque on our teeth but no acid forms. It may even help fight the bacteria by loosening their grip on the tooth enamel.

The inability of xylitol to cause tooth decay was demonstrated in a large-scale test on 11 and 12-year old children in a rural community in northern Finland, and was carried out by the Ylivieska Health Care Centre from 1982 to 1984. Those who chewed xylitol gum three times a day, taking in about 10 g of xylitol, had half the number of dental cavities as a group who were not given the gum. When the children were examined again in 1987 those who had chewed xylitol gum still had 50% fewer cavities.

If you put a xylitol sweet in your mouth you may even recognise it in a rather unusual way, by a brief sensation of coolness on the tip of your tongue. You are not imagining things: you are experiencing the chemistry of xylitol. When this simple sugar dissolves in water it absorbs heat and the solution gets cooler. The same happens on your tongue. Xylitol is slowly gaining a market as a sweetener for chewing gum, for sweets that you suck, like mints and fruit gums, and for pharmaceutical tablets that are to be kept in the mouth such as throat lozenges, decongestants, and pastilles used to heal mouth ulcers.

The old name for xylitol is birch-sugar, and most of it still comes from Finland where there are whole forests of birch trees. Xylitol is made by treating birch wood with water, but the processes of extraction are relatively costly. This is why xylitol is not a cheap substitute for sugar if a formulation requires a lot, but it makes an ideal coating for things like chewing gum and medicinal products.

Of course xylitol was not without its critics, and they were alarmed when investigations in the 1970s showed that it caused various disorders in rats. However, these effects were only observed when the rats were fed abnormally large amounts of xylitol, and this sweetener was never at risk of not being accepted by the licensing authorities because it occurs naturally in several fruits and vegetables. Strawberries and raspberries are particularly rich in xylitol, and so is cauliflower. Both yellow plums and greengages contain almost 1% of xylitol.

It is not economical to extract xylitol from any of these sources, but it can be got from wood. Wood consists of three main components: cellulose, which is used to make paper; lignin, which consists of polyphenols and has few uses; and hemi-cellulose, which is like cellulose but is a polymer of other saccharides, and is mainly xylose from which xylitol is made. Hemi-cellulose rich in xylose is also part of other plants such as corn cobs, cotton seed, sugar-cane bagasse, straw and nut shells. The xylose can comprise up to a quarter of the dry weight of these materials. Xylose is easier to extract than cellulose, although the process to turn it into xylitol requires several stages; the end product is pure crystals of xylitol.

Xylitol may solve the problem of tooth decay, but it still has the calorific value of a carbohydrate-like sugar. There are some bulk sweeteners which get round this problem, and **erythritol** is one. This is found in fruits, fermented foods and soyabean sauce. It has 75% the sweetness of sugar, but no calories, nor does it cause tooth decay. Erythritol is rather like xylitol but with one fewer carbon atom. It was discovered in 1874, and it can be made from glucose by fermenting with the right yeasts, but the yields are low and the processes are not yet commercial. Currently the Japanese Nikken Chemical Company of Tokyo are investigating a fermentation process which will convert glucose into erythritol with almost 50% yields and with few side-products.

Artificial sweeteners

The alternatives to bulk sweeteners are intense sweeteners, of which a few milligrams will give the same response on the tongue as a gram of sugar. Not all of these are artificial, although the most successful ones are. Because such molecules are the product of the chemical laboratory, they are automatically viewed with suspicion by some. Even so, millions of us willingly consume them every day.

The term 'artificial sweetener' has been much abused and carries such tones of disapproval that those in the industry prefer to use the phrase 'non-nutritive sweetener', which may be technically correct but has little else to recommend it. Intense sweeteners are big money earners with sales exceeding $1 billion per year. The table below lists several of them, divided into the truly artificial ones and those that come from plant material.

Sweetness factors (compared to sucrose taken as 1)	
artificial sweeteners	
cyclamate	30
aspartame	180
acesulfame	200
saccharin	300
sucralose	600
alitame	2000
plant-extract sweeteners	
steviosin	300
hernandulcin	1000
neohesperidine	1800
thaumatin	2000

The first artificial sweetener

Before we look more closely at modern day intense sweeteners, it will be instructive to look at an earlier one, *sapa*, which should definitely have carried a health warning. *Sapa* was the artificial sweetener of Ancient Rome, and a part of civilised living at the time of Christ. It has a strange story to tell which shows the extent to which the demand for sweetness will often override other considerations.

Honey alone was never capable of satisfying the demand for sweet-tasting food, and in Imperial Rome sugar was little more than a rare novelty from the East. *Sapa*, on the other hand was easy to make, it found a ready outlet in the sweetening and preserving of wine, and was especially popular with cooks for making sauces. Roman cooks were very fond of *sapa*, and it is mentioned in 85 recipes of the Roman Empire's best-seller, *The Apician Cookbook*. *Sapa* was like honey, an intensely sweet syrup, but the dangers of using it were great indeed.

Wine producers also used *sapa* to improve the quality of wines and extend their life. The Greeks discovered this useful little ruse, and although Greek wines were much admired they also had a reputation for causing miscarriages. Pure *sapa* was eaten neat by Roman prostitutes deliberately to produce abortions, and for another reason: it gave them an attractive pale complexion.

The Romans suspected *sapa* was deleterious to their health, but they continued to use it. It was reputed to make people tired and listless, anaemic, constipated and infertile. It did all this because it was made from lead. Pliny the Elder, who lived from 23 AD to 79 AD, described how *sapa* was prepared by boiling down grape juice, wine lees, or sour wine in pans until most of the water and alcohol had gone. If you want good *sapa* he wrote, then it was essential to use a lead pan.

All can be explained with a little bit of chemistry. Wine contains natural organic acids, such as tartaric and citric acid, and if wine picks up spores of the bacterium *Acetobacter*, which are quite common, then it will become more acidic as the alcohol is converted to **acetic acid**. Eventually it becomes sour wine, *vin egar*, i.e. vinegar. The Romans did not waste wine that was going off, but made *sapa* from it. Lead-lined pans were essential for this purpose since the lead was attacked and dissolved by these various acids which reacted to form lead salts, mainly **lead acetate**. The lead compounds were responsible for the sweet taste, and modern analysts have demonstrated that *sapa* would have had 1000 parts per million of lead (0.1%). Beautiful white crystals of lead salts could be grown from *sapa*, and these tasted very sweet. When their chemical nature was realised in the 18th century they were called sugar of lead.

Sapa did all that it was reputed to do. It sweetened food, it preserved wine by killing bacteria, it caused abortions because of the lead, and it produced pale complexions because a feature of chronic lead poisoning is anaemia. What the Romans did not realise, however, were other side effects of a high lead intake, which are constipation, stomach pains, muscular weakness and headaches. The reputed infertility of the upper classes is now thought to have been due to *sapa*. This artificial sweetener passed into history along with the Roman Empire, but not the trick of putting lead into wine. Even in the last century vintners were still adding lead to port wine, and the traditional trick of the trade was to put a pellet of lead shot in every bottle.

Saccharin

In the years when sugar was cheap and plentiful there was no demand for artificial sweeteners. When the first modern sweetener was stumbled upon in the last century there was already a need building up, and one which kept up a demand for it. Doctors who told their patients to eat less sugar no longer condemned them to a life without sweetness. They could tell them to choose saccharin to flavour their drinks and desserts instead.

Saccharin was discovered by Ira Remsen and Constantin Fahlberg at the Johns Hopkins University in the USA in 1879. They were researching sulfur–nitrogen compounds and made a new derivative of benzene. In those days it was common practice to record the taste of all new substances and the intense sweetness of the new compound was reported. Its taste led to the name saccharin.

Saccharin is over 300 times sweeter than sugar, and a tiny tablet of saccharin is as sweet as a spoonful of sugar but without its 20 Calories. It does not taste quite as nice as sugar, and some people say it has a metallic aftertaste. Its safety had been debated endlessly, and saccharin is probably the most investigated ingredient in the food industry.

During the last century there were almost no tests for health and safety before marketing a new food additive, and manufacturers were quick to realise the potential of saccharin. Even so, tests were soon demanded when they started to use it as a cheap substitute for sugar. Testing in those days consisted of the blunt but effective method of feeding 5 grams of saccharin a day for six months to volunteers. They came to no harm and public fears were allayed.

Some people continued to oppose it even so. Controversy raged and the then US President Theodore Roosevelt set up a review board, which concluded that saccharin was safe below 0.3 g per day. Saccharin became part of the regular diet of diabetics and the occasional diet of slimmers. When sugar rationing was introduced in both World Wars, saccharin was used by millions of people. Manufacturers of fruit drinks found that a blend of saccharin and sugar gave a better, less syrupy product than sweetening with sugar alone. In this way saccharin became an unsuspected part of many people's diet. In the 1950s, when the modern vogue for slimness began, saccharin became popular in its own right, and its use was not challenged again until another artificial sweetener, **cyclamate**, was banned in the USA in the 1970s.

Saccharin was suspected of causing bladder cancer because of its similarity to cyclamate. The two were often used in combination. When saccharin was tested on rats, a few male rats, but only slightly more than expected, developed bladder cancer when fed extremely high doses of saccharin (up to 3% of their diet) throughout their lives. One observer said it was equivalent to drinking 800 cans of diet cola per day. The evidence is difficult to translate into human terms. Saccharin had been around since great grandmother's day and if it had been causing bladder cancer this had gone unnoticed. Saccharin was not banned, and since then many other studies have failed to find any link between saccharin use and human cancers. Saccharin passes unchanged through the body and does not react with DNA, which is what has to happen with a cancer-triggering chemical. Saccharin is still approved for use in 90 countries around the world. It cannot promote dental caries and there is the suggestion that it may even help prevent this condition because it inhibits the micro-organisms in plaque by interfering with their cellular membrane.

Cyclamate

Cyclamate is also a sulfur–nitrogen compound, and again it was pure chance that led to its discovery as a sweetener in 1937. When Michael Sveda of the chemical company E.I. DuPont de Nemours put his cigarette down on the edge of his laboratory bench, it picked up a few grains of the material he had just made. When he next put it to his lips he was struck by an intense sweet taste, and cyclamate had been discovered. This was to be an even bigger money-spinner than saccharin, because it had three advantages: it was patentable; it did not have a bitter aftertaste; and you needed more of it to get the same level of sweetness because it has only a tenth of the sweetness of saccharin. However, you still needed only one thirtieth of the amount of sugar. By the mid-1960s cyclamate dominated the US market for artificial sweeteners.

The warning bell for cyclamate had sounded in 1957 when a research team in England found that mice exposed to saccharin developed bladder cancer. They had observed this after they had implanted a pellet directly into the animals' bladders, a somewhat unusual method of testing. This evidence was not accepted by the FDA as indicative of a risk to humans, but the seeds of doubt had been sown. A later study showed that the same thing happened with a mixture of cyclamate and saccharin. A lifetime study was undertaken in which rats were fed high dietary doses of cyclamate, saccharin and cyclohexamine. This last chemical is the compound that cyclamate turns into inside the body, and was also suspect. The study found an increased number of bladder tumours in male rats. A media blitz against cyclamate was unleashed and the FDA banned it. Cyclamate had fallen foul of the 'Delaney clause' of the US 1958 Food Additives Amendment which said: 'no additive shall be deemed to be safe if it is found to induce cancer when ingested by man or animal.' The Canadian and UK governments also outlawed cyclamate, but other countries were unimpressed by the tests because of the abnormally high amounts fed to the animals, and cyclamate is still popular in many countries around the world. Then in 1984 the FDA and the National Academy of Science announced that cyclamate was not carcinogenic after all. But by then it was too late to reinstate it as the leading sweetener in the world's largest market.

Acesulfan

Acesulfam is the latest of the sulfur–nitrogen sweeteners. It is about 200 times sweeter than sugar, and was discovered in 1967 by Karl Claus of the German chemical company Hoechst, again quite by chance. Claus licked his finger to pick up a piece of paper in the lab and got a pleasant surprise since it had picked up a trace of the new compound he was working on.

Aware of the fate of the other sulfur–nitrogen sweeteners, Hoechst tested acesulfam very thoroughly before they put it on the market. Acesulfam, which is sometimes called **acesulfam-K** (K is the chemical symbol for potassium, and acesulfam is a potassium salt), goes by the trade name of Sunett and it is used in soft drinks, desserts and puddings, as well as being sold as individual tablets for sweetening tea and coffee. It is approved in over 50 countries, including the USA, and there are over 2000 products throughout the world that contain it, including toothpastes, mouthwashes and pharmaceuticals. This sweetener has a synergistic effect with the next one we are going to look at, **aspartame**, which means that each boosts the sweetness of the other. Used in combination, much less of either is required to achieve the same intensity of taste.

Aspartame

Most people will probably know aspartame by its brand name of NutraSweet, and products such as Canderel or Equal. Sales of this around the world now exceed $1 billion, and in the US this artificial sweetener outsells all others. It was discovered in 1965 by James Schlatter working for the US firm G.D. Searle & Co. of Stokie, Illinois. Schlatter was making compounds for testing as anti-ulcer drugs, and had joined two protein amino acids together, **aspartic acid** and **phenylalanine**. He discovered that the product tasted sweet, but the company found it even sweeter when eventually they sold out to the chemical giant Monsanto for $2.7 billion in 1986. Today there are manufacturing plants in the USA, Japan, Brazil and France.

While it was protected under patent, Searle was able to charge a very high price for NutraSweet, about $100 per kilogram, and people were prepared to pay the price because so little aspartame was required. It is 200 times sweeter than sugar so that only a few milligrams were needed to flavour a cup of tea or coffee. Nor only that, but it tasted as good as sugar, and some even preferred its 'clean' taste to that of sugar.

Like all artificial sweeteners, however, it had its critics. The FDA approved aspartame in 1974, then had second thoughts in 1975 and demanded more stringent tests. They finally cleared it for general use in

1981 provided it carried a warning that it contained phenylalanine. Now at first sight this seems rather strange because phenylalanine is an *essential* amino acid—in other words we cannot live without it. Nevertheless the warning was necessary for those people who suffer from a condition called phenylketonuria. They are sensitive to excess phenylalanine because they lack the enzyme which disposes of it when there is too much of it in the body. About one person in 15 000 is affected by phenylketonuria, and they need professional advice from a hospital dietitian. They must monitor what they eat and avoid foods that have a lot of phenylalanine, like steak, eggs, cheese, yoghurt and chocolate. Knowing that NutraSweet also contains phenylalanine means that they should avoid this artificial sweetener.

Aspartame is digested in the human body, as we might expect from its being a tiny protein fragment. It breaks down into aspartic acid, phenylalanine and **methanol**. The first two are amino acids which our body may use as building blocks in making its own protein. The amount of aspartic acid and phenylalanine we get from aspartame is tiny compared to the amount in a normal diet where they come from foods such as meat, milk and fish. The methanol has to be disposed of—as we shall see in Chapter 3, this is an undesirable molecule to have in our body—but again the amount we get from aspartame poses no risk. A glass of fruit juice, beer or wine will give us just as much.

Attacks on aspartame came from several quarters. Some claimed it raised the level of phenylalanine in the brain, which might affect those suffering from high blood pressure and Parkinson's disease. In 1984 the US public interest group, Common Cause, complained that the testing of aspartame by the FDA had been seriously flawed. In the UK, the national newspaper the *Guardian* carried several articles in 1983, the year when NutraSweet was launched in the UK, that questioned the safety of aspartame. They reported the claims by Professor Richard Wurtman of the Massachusetts Institute of Technology that aspartame interferes with the body's chemical messengers, increasing the desire for food, affecting blood pressure, making us more sensitive to pain, and causing insomnia.

The anti-aspartame campaigners kept up their attacks throughout the 1980s with claims that it even caused eye damage and mental illness (that came at a meeting of the American Association for the Advancement of Science in 1988). In Britain the *Guardian* again attacked NutraSweet in July 1990, making more serious allegations in which they implied that the manufacturers had faked laboratory tests submitted to the FDA when they applied for approval of aspartame as a sweetener. NutraSweet sued the *Guardian* for libel and malicious falsehood and won substantial damages.

The anti-aspartame campaigners did not have it all their own way, and there were even reports by a group at the University of Illinois in Chicago that the sweetener protected against tooth decay. Research at the University of Leeds, England, showed that it helped dieters by reducing their appetite slightly. Such reports generally failed to get a public airing beyond specialist science journals and trade magazines.

The makers of NutraSweet were confident that they had a winner, and knew that on chemical grounds it was just like a tiny piece of meat as far as the body was concerned. It now dominates the diet drinks and desserts market. For 99.994% of the population NutraSweet can be of real benefit. The unlucky 0.006% who suffer from phenylketonuria must use other artificial sweeteners. Despite all the attacks, NutraSweet has survived and it is used in 5000 products in almost every country of the world.

The patent for aspartame expired in 1992 and other companies can now challenge NutraSweet, whose production is about 10 000 tonnes a year and growing. With healthy competition the price should fall for what is after all a relatively cheap chemical. The present makers will no doubt dominate the market since they have exclusive rights to the brand name NutraSweet, which they have carefully fostered.

There are still two important features of an ideal sweetener that aspartame lacks: stability to heat and stability in solution. You cannot bake with the raw powder and this is a major drawback, although 'encapsulating' aspartame to protect it is seen as one way round this problem. Its stability in solution is less of a problem, even though a few per cent of the aspartame in a drink decomposes each month to a compound with no taste, and it decomposes more quickly if heated. It is reasonably stable in acid drinks like colas, or lime and lemon drinks, and at low pH 4 it has a shelf-life of six months or more.[9]

[9] The measure of acidity is the pH scale which ranges from 0 (very acidic) through 7 (neutral) to 14 (very alkaline). The acids we encounter in our everyday life range from weakly acid rain water, which has pH 5, to tomato juice pH 4, vinegar pH 3, lemon juice pH 2, and stomach acid pH 1.

Aspartame is made from two amino acids. Other combinations of amino acids are also very sweet, such as alanine and aminomalonine, and when this combination was discovered in 1985 it was claimed to be more stable than aspartame in water. Pfizer, the pharmaceutical company, has a sweetener which is a combination of aspartic acid and alanine under the name of alitame. It is 12 times sweeter than aspartame and much more stable. Normally small changes to a sweet-tasting molecule lead to large changes in taste, and tinkering with one of the amino acids of aspartame itself might have been expected to kill the sweet taste. Yet alitame still tastes as sweet even when one of its component amino acids is attached to the molecule back-to-front.

Sucralose

Sucralose is an artificial sweetener that will withstand heating, and it is water-soluble and stable in acids. Indeed, it comes near to being the perfect sweetener. It was discovered accidentally in 1976 at Queen Elizabeth College, London, by a graduate researcher Shashikant Phadnis who was reacting ordinary sugar with a reagent that attached chlorine atoms to the molecule. When he was rung up by Tate & Lyle, the sugar company he was collaborating with, and asked if they could have some samples of the material he was making for testing, Phadnis misheard them and thought they wanted it for *tasting*, and so tried it himself. The result was sucralose.

Sucralose has three chlorine atoms attached to a sugar molecule, and this multiplies the sweetness factor by about 600. Replacing more than three can give even more intense sweeteners, but these are difficult to make. Testing so far has shown this chlorinated sugar to be safe, but there is a snag. Most sucralose is taken in by mouth and passes through the body unchanged, but a tiny amount is split into its two components, *chloro*glucose and *chloro*fructose, and these are intrinsically dangerous and toxic, although not at the levels at which Sucralose would be used. The danger was enough to cause the FDA in 1992 to question the wisdom of using Sucralose, and to ask for more tests. Nevertheless, it has already been approved for use in Canada, Russia and Australia.

Although most intense sweeteners in common use are artificial ones, there are some that are derived from natural sources. These are given in the Box below.

Natural intense sweeteners

Thaumatin

This is a very large protein-like molecule extracted from the West African fruit ketemfe (*Thaumatococcus danielli*) and is 3000 times sweeter than sugar. Like aspartame it is compounded of amino acids, but instead of only two units it has hundreds of these joined end-to-end. It tastes so sweet because it can trigger many receptors on our tongue at once, but this also means it has the drawback of clinging to the tongue and the sweet sensation lingers too long. This may be used to advantage in some things such as chewing gum or bitter medicines, and it is also used in toothpastes, mouthwashes, jam and soy sauce. Animals love thaumatin and dogs and cats can be persuaded to eat canned pet foods, and pigs to put on more weight, if their food is flavoured with it. Thaumatin is sold under the trade name of Talin.

Stevioside

This is a mixture of saccharide molecules which come from a plant of the chrysanthemum family called *Stevia rebaudiana*, which was originally grown in Paraguay and used for centuries by the natives there to sweeten their food and drink. It is 300 times as sweet as sugar. It can be separated into various components called rebaudisodes which make up to 3% of the plant leaves, and these are even sweeter, but they have a lingering bitter aftertaste. Stevioside has the advantage that it is stable to heat. It has been used in South American and Japan for many years and it does not cause tooth decay.

Glycyrrhizin

This comes from the liquorice root and is more than 50 times sweeter than sugar. It also has several medicinal properties, and research has shown that it is an anti-ulcer, anti-inflammatory, and anti-spasmodic agent. There are even claims by its supporters that it is an anti-viral agent and good for the heart.

(continued)

Natural intense sweeteners (*continued*)

Neohesperidine

This is derived from a naturally occurring chemical found in orange peel. At concentrations less than 4 parts per million (ppm) it is not perceived as sweet, but it enhances fruit flavours and suppresses bitter tastes. At higher concentrations, above 5 ppm, it registers as several hundred times sweeter than sucrose. It enhances the sweetness of other sweeteners and allows the amount of these to be reduced. Neohesperidine lingers for a long time on the tongue and has a liquorice aftertaste. It has been approved for use in Europe, and is used in chewing gums and in some speciality Belgian beers.

What makes a molecule taste sweet?

Most of us would consider flavours to be the special preserve of cooks and chefs, and to look upon their blending as an art, not a science. Yet flavour is part of food chemistry, and, just as with fragrances, there is an industry whose job it is to extract long-established flavours, analyse them, and then to make them by chemical means. Some chemists also research into new flavours, hoping to discover and prepare entirely different tastes for our delight. The chemistry of taste may still be in its infancy, but great strides have been made in the past few years in understanding at least one area of taste, and that is sweetness. Chemists now know enough about the nature, or rather the 'structure', of sweetness to be able to predict which molecules will produce this sensation on our tongue.

So what is the secret of sweetness? The answer does not lie, as you might expect, with the chemical behaviour of a molecule, and if you search for a common feature among those materials which taste sweet you will find none. We can see this if we compare sweeteners like lead acetate, sucrose, aspartame and saccharin which are all chemically very different. Yet surprisingly the answer to sweetness lies with their molecular structure, or at least to a triangular part of it.

Just how subtle taste can be is demonstrated by the amino acid phenylalanine, which in its right-hand form tastes slightly sweet.[10] When you make the left-hand form of phenylalanine, you find it

---

[10] Left and right-handed molecules are discussed in Chapter 5.

tastes bitter. These two molecules are exactly the same chemically, with the same atoms and groups joined by exactly the same chemical bonds. The difference is merely that they are mirror images of each other, and this is enough to alter radically their taste. Our own experience of putting on shoes the wrong way round tells us that an apparently small difference can have disproportionate consequences when we mis-match left- and right-hand objects, and so it is with receptors on our tongue. Just as a key only needs a small modification for it not to turn a lock, so a molecule with the wrong sized sweetness key will not turn on the sensors in our taste buds.

In the 1960s Robert Shallenberger and Terry Acree suggested that molecules could activate the sweet buds on our tongue if the hydrogen atoms they contained were positioned at exactly the right angle to lock on to atoms in the receptor. Lemont Kier of Massachusetts College of Pharmacy at Boston identified three parts of a sweet molecule that determined the shape to fit the receptor, and this became known as the triangle-of-sweetness theory. Support for this suggestion was forthcoming when molecules were constructed with the shape he predicted, and were shown to be sweet.

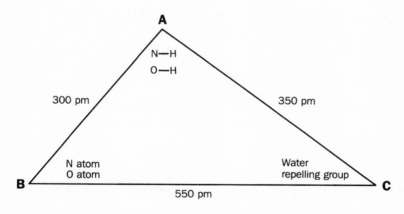

The triangle of sweetness. If a molecule is to taste sweet then it must trigger the sweetness receptors on the tongue. To do this it requires three features: an oxygen or nitrogen atom carrying a hydrogen (OH or NH) at location A; an oxygen or nitrogen atom at location B; and a water repelling group, such as a hydrocarbon, at location C. The numbers along the sides of the triangle are the ideal distances between these three centres expressed in picometres (10^{-12} metres).

At one corner of the triangle of sweetness there has to be a hydrogen atom that will lock on to oxygen or nitrogen on the receptor. In so doing it forms what is called a hydrogen bond. This is a weak interaction between molecules and relies on the attraction between a hydrogen atom (which is positively charged) in one molecule, and an oxygen or nitrogen atom (which is negatively charged) in another molecule. At the second corner of the triangle of sweetness there has to be an oxygen or nitrogen which will similarly attract a hydrogen on the receptor, again forming a hydrogen bond. At the third corner there are hydrogens that do not want to engage in hydrogen bonding, and these are hydrogens attached to carbon atoms. Get all these components in just the right positions and you can almost guarantee the molecule will taste sweet. The triangle of sweetness fitting a receptor is rather like an electric plug fitting a socket; when molecule and receptor connect a current flows to the brain and we sense sweetness.

Even though we can design a molecule which will be sweet, there are other factors that may make it unsuitable for commercial use. Most of the artificial sweeteners discovered so far decompose when heated, which is a big disadvantage because it excludes their use in cooking. Strangely, the first ever sweetener, *sapa*, did not suffer from this defect, which explains why it was so popular with the chefs of the Roman Empire. Sugar is still the only chemical that retains, and even improves, its sweetness on heating. What the world needs now is an artificial sweetener we can cook with, so that we can enjoy reduced-calorie cakes, biscuits, fruit pies, tarts and candies. Sugar makes up at least a third of these, and in some cases they are almost pure sugar.

The perfect sweetener has yet to be discovered or designed, although we can suggest its specifications: it should be hundreds of times sweeter than sugar yet with the same 'rounded' flavour and taste profile, in other words quick to register on our receptors and just as quick to de-register. Nor do we want any aftertaste like the metallic bitterness that some people experience with saccharin. Our model sweetener must of course be non-toxic (a child might eat a whole packet at once), and it must not cause dental caries. Either it has to pass straight through our gut or, if it is absorbed, it has to be digested like other components of our food. The manufacturers' requirements should not be overlooked either. They want it to be cheap to make and easy to purify, and to have a long shelf-life. It should be soluble in water and alcohol, unaffected by acids, and stable when heated at least to normal cooking temperatures. Find such a molecule and your fortune is made.

We know that the perfect sweetener will not solve the problems of the overweight people in Western societies, and if past campaigns against artificial sweeteners is anything to go by, there will always be someone ready to accuse it of being a health risk. The irony of low-calorie sweeteners is that they have not led to a decrease in the numbers of overweight people, but this clearly is a social problem and not one that can be solved by making better sweeteners. Indeed, sweeteners may even be counterproductive if they encourage people to eat more of so-called 'diet' foods. These people are using the skills of the chemists to indulge their longing for sweet-tasting foods.

The ever-increasing sales of convenience foods keeps the sugar producers in business despite the apparent decline in the sales of sugar, jam, golden syrup and all the products our mothers bought to bake the traditional home-made treats of cakes and pastries. Artificial sweeteners can only help people if they use them to *replace* some of the sugar in their diet, and thereby keep their weight under control. Sweetness, whether natural or artificial, bulk or intense, is not something to fear. A little chemistry and a little care, and you can enjoy this pleasant sensation for the rest of your life, which may even be longer and healthier, thanks to artificial sweeteners. These chemicals can promise nothing more.

CHAPTER THREE

Alcohol

Alcohol and body chemistry; the production and chemistry of alcoholic drinks; fraudulent drinks; the risks and benefits of drinking alcohol; industrial alcohol.

WHEN sugars ferment under the action of yeasts they generally produce alcohol. Unlike sugar, for which chemists can make artificial alternatives as we saw in Chapter 2, there is no substitute for alcohol. In this chapter we will take a chemist's eye view of this molecule—how it forms, what effects it has on us, how it offers us an alternative fuel, and how we can learn to live with it and not die from it. We all have alcohol naturally in our body in small amounts, and it has no effect because there is so little of it. When we suddenly increase the amount, by drinking a lot of it, then we experience some rather unusual effects—elation to begin with, but deflation a few hours later. Were alcohol to be discovered today its sale to the public would never be permitted because of the potentially lethal side-effects.

There are many people in the world for whom alcohol is taboo, such as Muslims and Mormons. Muslims are forbidden it supposedly because of Mohammed's displeasure when he discovered his troops were drunk on the eve of battle. A Muslim can only drink alcohol if it is prescribed by a doctor. Mormons and other Christian sects can find plenty of reasons for not consuming alcohol: it is addictive; it causes illness; it ruins lives; it leads to violent behaviour in young people; and it is a major cause of death on the roads. Alcohol has another, although less socially destructive, drawback: weight for weight it provides almost as much energy as fat: 7 Calories per gram compared to 9 Calories per gram for fats.

There are also those who suspect alcoholic drinks of harbouring other unwanted chemicals, and they claim that brewers, vintners and distillers use additives to improve the colour, the clarity and the keeping quality of their drink. And they do. More worrying still are reports of fraudulent drinks, made from cheap supplies of industrial alcohol to which other chemicals have been added to provide flavour and colour. If they are made with methyl alcohol, such drinks can kill. And they do.

Yet for every person who opposes the demon drink there is an ardent supporter of alcohol, and they too have good reasons for their opinions. Alcohol is enjoyable and relaxing; it makes social and sexual contacts easier because it releases us from our inhibitions; and there is evidence that those who drink alcohol suffer less from coronary heart disease, and have fewer colds. We may also endorse alcohol not as a drink but as a renewable and sustainable source of fuel for transport, and as a feedstock for the chemicals industry. It can be made easily and cheaply, it can be stored and transported safely, and it burns cleanly leaving little environmental pollution.

It is rare to find such a collection of emotions, warnings, hazards and benefits surrounding a simple molecule. So what is special about this chemical? Can we balance its advantages against its disadvantages? Can we prevent addiction to it? Can we detect phoney drinks? The answer to these questions is to be found in a closer look at its chemistry.

Alcohol is also called ethyl alcohol or **ethanol**,[1] which is the correct chemical name for this molecule.[2] There are two ways in which alcohol is produced on a large scale: by the fermentation of sugars, or by the chemical reaction of water and **ethylene**. The former is agriculturally derived alcohol; the latter is often referred to as industrial alcohol. Although they are chemically identical, agriculturally derived alcohol is classed as drinkable, while industrial alcohol is made undrinkable by adding unpleasant chemicals. It is then popularly referred to as methylated spirits or meths for short.

---

[1] A reminder that when the name of a substance appears in **bold** type, it means that there is more about its chemistry in the Appendix.

[2] The name alcohol comes from the Arabic *al kohl*, but its origin is not clear.

The alcohol which is made by the fermentation of agricultural products is produced by the activity of fungi called yeasts. These grow by feeding on sugars such as glucose, and as they do this they make alcohol, which for them is an unwanted by-product. Alcohol is a small part of the cycling of carbon through the environment, a subject we shall return to in Chapter 9. We could follow an individual carbon atom through this cycle and the scenario might go as follows. The carbon begins as carbon dioxide in the atmosphere, which is absorbed by a leaf of barley and turned into carbohydrate. After the barley has been harvested, and malted to break the carbohydrate down into simple sugars, it is then digested by yeasts whose enzymes turn it into alcohol. It may then be drunk. Inside the drinker the alcohol is processed and its energy extracted, thereby turning it back into carbon dioxide. This is breathed out and returned to the atmosphere. There are two halves of such a cycle, production and consumption, and since the latter affects us intimately when we drink alcohol I will begin with the chemistry of this.

Alcohol and body chemistry

We can analyse the contents of an alcoholic brew and find that it contains lots of things apart from alcohol and water. For instance a litre (about two of the traditional pints) of the world-famous Irish drink, Guinness, in addition to providing you with 370 Calories of energy, contains the following:

alcohol	30 g	thiamine (vitamin B_1)	0.04 mg
carbohydrate	30 g	riboflavin (vitamin B_2)	0.4 mg
protein	3 g	pyridoxine (vitamin B_6)	0.6 mg
phosphorus	557 mg	nicotinic acid (niacin)	6 mg
iron	0.3 mg	biotin	0.01 mg
sodium	25 mg	folic acid	0.07 mg
potassium	300 mg		
calcium	54 mg		
magnesium	133 mg		

Clearly Guinness is also a good source of some essential minerals, and if we drink it regularly it will provide a significant fraction of our daily requirements of the B vitamins. The surprise of drinks like Guinness is the large amount of potassium compared to sodium, and this is true of all beers. Most foods have much more sodium than potassium.[3]

Despite these dietary components an alcoholic drink is not regarded as a food, a medicine, or a tonic, although in earlier times alcohol was advertised as all of these. Today we treat it mainly as a relaxant. Our body treats it as a poison. There are many ethical questions that alcohol raises which a knowledge of its chemistry cannot answer, but it can answer two key questions. Why does drinking make us feel good for a while, and then a few hours later make us feel so bad? Why do many people have to take special care because their tolerance of alcohol is low, such as women, native Americans and Japanese?

Alcohol does not occur as part of the complex chemistry going on inside the cells of the human body, but a little alcohol is produced naturally within our intestines by bacteria and yeasts which have enzymes that can turn carbohydrate into alcohol, and this alcohol gets into our bloodstream. Our liver has an enzyme called alcohol dehydrogenase (ADH) with which to get rid of this unwanted chemical, and it converts the alcohol to **acetaldehyde**. This molecule is then turned into **acetic acid** by another enzyme, and the acetic acid is used in the so-called Krebs cycle of chemical reactions, by which its energy is extracted and it emerges as carbon dioxide.

The liver can deal with large amounts of alcohol, but it needs time to do so. If a person takes in too much alcohol at one session they may even die, and a lethal dose can be as little as 400 ml, the amount of alcohol in a litre bottle of spirits. A normal liver can only process 12 ml (10 grams) of alcohol per hour. A glass of wine, which is about 12% alcohol, may provide this amount, as will half a pint of beer which is around 4% alcohol, and so will a tot, or shot, of spirits which is generally 40% alcohol. These measures are only rough guides because the amount of alcohol in wine, beer and spirits can vary within quite wide margins. The quantity commonly referred to by those who give advice about drinking is the 'unit', which is generally that quantity of a drink which will deliver 10 grams of alcohol. It is not a precise amount, only a guide, since it varies widely even among drinks of apparently the same type.

---

[3] We often redress the balance when drinking by eating salted snacks.

The effects of alcohol

Units	Number of drinks			Alcohol level in blood (mg per 100 ml)	Effects on the average man*
	beer	wine	spirits		
2	1 pint	2 glasses	2 singles	30	feeling of well-being
3	1.5 pints	3 glasses	3 singles	50	lack of inhibitions
5	2.5 pints	5 glasses	5 singles	80	unfit to drive
6	3 pints	1 bottle**	6 singles	100	unsteady on feet
10	5 pints	1 litre	10 singles	150	slurring of speech
12	6 pints	2 bottles	$\frac{1}{2}$ bottle***	200	drowsy and confused
18	9 pints	3 bottles	$\frac{3}{4}$ bottle	300	drunken stupor
24	12 pints	4 bottles	1 bottle	400	dead drunk, maybe dying

*For a woman the same effects may be experienced with about two thirds the amount of alcohol that a man needs, assuming they are the same weight.
**Wines are traditionally sold in 700 ml bottles.
***Spirits are often sold in bottles of 700 ml, and this is what is meant by a bottle here, not the 1 litre bottle which is becoming more common.

The effects of drinking alcohol can be seen in the table above. This table cannot take into account all factors which determine the effect that alcohol will have, because this depends on a person's weight, the condition of their liver, and whether the alcohol is taken with a meal.

When we drink alcohol it starts to be absorbed the minute it enters our mouth, but most is absorbed from the stomach and the gut. Ten per cent of the alcohol we take in is lost from our body without being digested. It is exhaled on our breath, sweated through our pores, or passed out in our urine. The other 90% is processed by the liver. The two chemicals which most affect our body when we drink are alcohol itself and acetaldehyde, and these affect different organs in different ways.

Brain and central nervous system: alcohol makes us feel happier, more at ease with people, and less inhibited about what we say and do. Technically alcohol is referred to as a depressant, but this does not mean that it makes us depressed. It means that it slows down the activity of the central nervous system, so that messages take longer to travel along nerve fibres. We become more relaxed and overconfident of our abilities, but also slower to react, and our speech becomes slurred. Alcohol has this effect on the brain because it replaces water molecules around nerve cells, and this interferes with the movement of electrically charged atoms which are responsible for transmitting information along a nerve fibre. Alcohol also slows the movement of chemical messenger molecules which carry information from cell to cell.

Ears: these are the organs which give us our sense of balance. Alcohol changes the density of the tissue and fluid in the ear, and the more alcohol we take the bigger are the changes, until we lose our normal sense of balance. The result is that we sway and stagger, trying to compensate for the feeling that we are about to fall over.

Skin: here it is the acetaldehyde which has an effect, dilating the blood vessels and making us feel hot. The dilation of blood vessels in the scalp and around the brain results eventually in a bad headache. Alcohol raises our pulse and blood pressure which increase the sensation of warmth. We can achieve the same effect if we take a shot of spirits on a cold day, but the comforting idea that people dying from exposure in the snow can be saved by a drink of brandy (carried by a St Bernard dog, of course), is a popular myth. In fact the alcohol would only serve to increase the loss of heat from their body.

Stomach: Men digest alcohol quicker than women because they have more ADH in their stomach, and this is the enzyme that converts alcohol to acetaldehyde. Consequently men can tolerate more alcohol than women, because drink-for-drink, less alcohol gets into a man's blood stream from his stomach. Native American Indians and Japanese people have the same lower level of ADH as women. The male and female kidney disposes of alcohol at the same rate, so that is not the reason for the difference between the sexes. Tests in the US showed that when both men and women were given the same amount of alcohol by injection into their bloodstream there were no sex-related differences in behaviour.[4]

---

[4] Even so, alcohol generally has a greater effect on women than on men because of the difference in body weight.

Liver: this is the main organ for removing alcohol, but as we have seen it does this only slowly. Indeed, people have been known to fail a breathalyser test 24 hours after their last drink. Alcohol also stimulates the breakdown of glycogen to glucose in the liver, so depleting the body's immediate store of energy. Although alcohol is high in Calories it does not satisfy your desire for food—indeed, it seems to stimulate it.

Kidneys: if you drink 250 ml of wine (about 2 glasses) you will lose at least 500 ml of water from your body as urine during the next two hours. Normally our kidneys will reabsorb and reuse water, and are prompted to do so by a hormone called vasopressin, which is released by the pituitary gland at the base of the brain. Alcohol reduces the amount of vasopressin, and so the kidneys fail to get the message to recycle water, which then passes to the bladder and out of the body. The result is dehydration unless this fluid loss is replaced.

Sex: alcohol can make us feel sexy and more easily aroused by lowering our inhibitions. It can prolong our performance in bed by slowing down the nervous system. Too much alcohol for a man can so relax his muscles that it interferes with his erection (the infamous 'brewers droop').

All these effects are the results of an immediate toxic reaction to alcohol. There are also long term effects of taking alcohol over many years, such as impaired brain function and memory. Excess alcohol can lead to acute inflammation of the stomach, especially if drunk on an empty stomach, and this is aggravated by the acetaldehyde formed, which is why for men one effect of alcohol can be a stomach ulcer.

The morning after, and how to avoid it

The body turns alcohol first to acetaldehyde, then to acetic acid, and finally to carbon dioxide. Acetaldehyde is the most likely cause of our hangover, because this is known to cause headaches and nausea. These unpleasant side effects can even be intensified, and this is one way to keep people off alcohol (see the Box on p. 68). Once alcohol is absorbed by our body we have its effects to contend with, and later those of its by-product acetaldehyde. As the alcohol disappears we sober up, but as the acetaldehyde forms we begin to experience the effects of this as a hangover. Sobering up is possible in theory by boosting the amount of ADH. Achieving this is another matter, but you cannot sober up by any of the popular remedies of sweating in a sauna, vigorous exercise, drinking black coffee or putting your head under the cold tap.

The morning after a bout of heavy drinking is simply the slow process of waiting for the acetaldehyde to disappear, but there are a few tips for coping with hangovers. No successful hangover cure has yet been marketed, and we know why this is unlikely to work for chemical reasons, unless it can speed up the removal of the excess acetaldehyde from the body. The following suggestions may relieve the worst excesses of a hangover:

- Have a glass of milk before you start drinking so you are not drinking on an empty stomach.
- Stick to one type of alcoholic drink, and occasionally have a soft drink.
- Drink a pint of water before going to bed.
- For breakfast eat something sweet, such as honey or jam. This is because these foods contain a lot of fructose, as we saw in Chapter 2, and this is the sugar that generates nicotinamide adenine dinucleotide (NAD) which is involved in the processing of alcohol. Extra sugar will also prevent depletion of the body's glycogen store.
- Avoid dark-coloured drinks such as port, sherry and red wines. However, the link between colour and hangover is not always a good guide. Whisky sometimes gets its dark rich colour from added **caramel**, and so does rum. But the reds of wines come from chemicals in the skins of grapes, and these may be partly responsible for a hangover.

The traditional remedy for a hangover is to resort to the hair of the dog that bit you and have another drink. This will only be of relief if you are addicted to alcohol, so that the unpleasant effects of your hangover are mainly withdrawal symptoms. Otherwise it is rather foolish advice.

People who are addicted to alcohol and really want to break their addiction can seek treatment and be helped to stop drinking. To prevent them succumbing to temptation and starting again, they can be prescribed a compound called **Antabuse** (see the Box below). This ensures that they will think twice before drinking any alcohol because the effects are so unpleasant. Antabuse works by blocking the enzyme which converts acetaldehyde to acetic acid, thereby removing it. A build-up of acetaldehyde produces a condition known as acetaldehydemia which can be extremely unpleasant. Even a little alcohol taken by someone on antabuse produces enough acetaldehyde for their body to react badly to it. They feel very ill because they are in effect experiencing a severe hangover, the symptoms of which are nausea, vomiting, laboured breathing, flushing, chest pains and throbbing headache. The experience is so dreadful that they will usually avoid alcohol while they remain on antabuse, although it has been found that some people can become tolerant to the drug and its effect is diminished. However, most people who take antabuse find it effective; but they must also be alert to the fact that some common household products contain alcohol, such as **vanilla** extract (35% alcohol), cough medicines (up to 25% alcohol), and mouthwashes (e.g. Listerine is 25% alcohol).

The Antabuse story

The effects of antabuse came to light in Denmark during World War II at a company called Medicinalco, where the policy was for employees to test any new drug that they discovered on themselves before it was prescribed for general use. And this is what Dr Jens Hald did while he was working on a drug against intestinal parasites such as worms. The chemical he took was called disulfiram, which had been used successfully to treat the parasitic skin disease scabies. Hald tested disulfiram on rabbits infected with parasites and it appeared to cure them. He then began to take the drug himself, and so did the company's medical research director, Dr Erik Jacobsen. They took a disulfiram pill every day for several weeks.

Most days they were fine, but on some days they went down with a bout of severe sickness, generally immediately after lunch. Curiously both men were only affected occasionally, and neither could link it to any particular cause—that is until one evening when Hald had a large drink of cognac with a friend. Immediately he became very ill while his friend was fine. He then realised that previous attacks coincided with meals at which he had drunk alcohol. When he told Jacobsen he too confirmed Hald's suspicion that it was disulfiram and alcohol that caused the vomiting and flushing. Tests on fellow workers confirmed that this was so. The immediate result was that disulfiram was dropped as a possible treatment for worms.

But that was not the end of the story. A few years later, in October 1947, Jacobsen was asked to stand in for another speaker at a public meeting and he told the disulfiram story as an amusing aside. This part of his talk was highlighted in a Copenhagen newspaper, the *Berlingske Tidende*, by a journalist who was at the meeting. The result was that alcoholics began writing to Jacobsen asking for some of his disulfiram pills. Clinical tests on alcoholics followed and a regime devised whereby Antabuse, the trade name Jacobsen gave the drug, could be used to help people to break their addiction to alcohol.

The production and chemistry of alcoholic drinks

Alcoholic drinks come in a large number of guises, but the main ones are beers, wines and spirits. What distinguishes the last of these from the other two is the chemical process of distillation which concentrates and purifies the alcohol. Concentrating alcohol this way led to a whole new range of drinks, but removing alcohol altogether requires more sophisticated chemistry. At the end of this chapter I will explain how this is done.

Grapes, barley, corn, sugarcane, fruit, potatoes and rice are all used to make alcoholic drinks. When the crop is harvested, and the sugars of the carbohydrate component are released, they may be fermented into alcohol by acting as food for a yeast. Yeast uses simple sugars in much the same way we do—as a source of energy—but unlike humans, who generate their energy with the help of oxygen from the air, the yeast generates its energy anaerobically, in other words without oxygen, and the product is alcohol.

Any source of sugary solution can be fermented, and in many countries there are drinks produced from a wide variety of fruit sugars. For example in the UK there is a large industry fermenting apple juice to make cider, and pear juice to make perry. There is even a cottage industry of homemade wine using soft fruit and wayside berries, such as blackberries and elderberries. Such fermented juice can be improved by adding ordinary sugar and the final strength of the drink may be as high as 13% alcohol.

All naturally fermented drinks contain hundreds of chemicals which come from the cereals, grapes or fruit. Other chemicals are also put into our drink, and additives commonly used by brewers and vintners are: **sorbic acid** and potassium sorbate, **sodium alginate**, ammonium caramel, sodium sulfite caramel, **silicon dioxide**, **propane-1,2-diol alginate**, and **sulfur dioxide**. All of these additives sound very chemical, but all have passed stringent health and safety tests. In fact some of them have been used for decades, and some, such as sulfur dioxide, have been used for thousands of years. As health risks they are insignificant compared to the risks from the alcohol itself.

Sorbic acid and potassium sorbate occur naturally in many fruits, and they are good at killing wild yeasts and other fungi. Sodium alginate enhances the texture of the drink by removing unwanted proteins. It is a long molecule made up of carbohydrate units joined together to form a chain, and is extracted from seaweed. Alginates are used to impart a smooth taste to many foods, such as ice cream. Ammonium and sodium sulfite caramels are forms of caramel and impart a rich golden or brown colour. Ammonium caramel is used mainly in beers, and sodium sulfite caramel in whisky and brandy. Silicon dioxide is a natural mineral that occurs in many forms, of which sand is the most common. It can be used in combination with isinglass, a natural gelatin, or as its sodium salt, sodium silicate, to clarify wine and beer by causing proteins, which may make these drinks cloudy, to cluster together and settle out. Propane-1,2-diol alginate is probably the most commonly used beer additive in the world; it produces a good frothy head. Sulfur dioxide is used in wines to destroy invading micro-organisms that might spoil the fermentation by producing other by-products.

In his well researched book *Name Your Poison* Ted Parratt discusses everything that is added to drinks, and makes the point that unlike foods, which must list all their ingredients, drinks need list only their alcohol content. All the drinker can do is trust the integrity of brewers, vintners and distillers. This is expecting rather a lot, in view of their behaviour down the ages, but with today's international news coverage the producers know how costly the wrong kind of publicity can be.

Wine

Wine can be made simply by crushing grapes, squeezing out the juice and allowing it to ferment with the wild yeasts that are naturally present. This can be a hit-and-miss affair, and does not always produce the desired result. It is better to add the right yeast to the grape juice, but even doing this still does not always stop millions of litres of wine going to waste because rogue yeasts have outpaced the starter yeast which the vintner added. Some rogue yeasts even produce toxins which can kill the commercial starter yeasts. Not all wild yeasts are bad: some were discovered to impart a new flavour to the wine, and so are cultured and added to the crushed grape. Winemakers jealously guard their strains of yeast. The preferred yeast for wine is *Saccharomyces ellipsoideus*, and unwanted bacteria are kept down by using sulfur dioxide, which is added as its solution in water or as sodium sulfite which dissolves in water and forms sulfur dioxide. In earlier times 'sulfiting' was done simply by burning sulfur near the vat of grape juice.

There are thousands of wines which divide first into red, white and rosé, then into region. White wines are made from grape juice which has been separated from the grape skins. Red wine is fermented with the skins and their colour chemicals are extracted into the alcohol solution. This also extracts **tannins** and other compounds which add flavour and help the wine to keep better. The red dye compounds from the skins can be extracted quicker by heating to 50 Celsius. Rosé wines are produced from a blend of white and red grapes.

The amount of alcohol in wines can vary between 5 and 13%, which is about the maximum concentration that a yeast can survive. Most have about 10–12%. If the content exceeds 13%, then the product must include added alcohol, and these are called fortified wines of which the best known are sherry, vermouth and port with about 15% alcohol. These also have added sugar (vermouths have added herbs as well), and they are left to stand in casks from whose wood they also extract more chemicals.

The world's major wine producers, in order of output starting with the largest, are: Italy, France, Spain, Argentina, the USA and Romania. Grape cultivation needs an average annual temperature of 10–12 °C and must not fall below 15 °C during the months April to October (September to March in the southern hemisphere). Traditionally the northern limit for grape cultivation is latitude 50°, which should exclude all of the British Isles, but there are a few vineyards in Southern England. Cultivation also depends on altitude, soil, and weather.

Vines bloom and form fruit, but until August these remain green and hard, with a high acid and a low sugar content. Throughout September things change rapidly as the fruit ripens, and the sugar content, which is mainly glucose, rises to over 200 grams per litre of grape juice. Ideally grapes should be picked at their peak glucose level because this is the sugar the yeast thrives on best.

The bunches of grapes are gently squeezed between rollers, so as not to crush the seeds or break the stems. These are then separated out with a stemming machine. The crushed grapes are pressed to release their juice which is called the 'must'. The harder the pressing the more juice is obtained, but the top quality juice comes off first. The average yield of must from a tonne of grapes is 750 litres. The must can be pasteurised to kill rogue yeasts by heating to near boiling (87 °C for two minutes) and then pure wine yeast has to be added. If the must has undesirable odours these can be removed by filtering the liquid through activated charcoal filters.

Fermentation takes about three weeks and is traditionally carried out in large oak vats, but more likely nowadays it is done in stainless steel or in glass-lined or plastic-lined tanks. White wines ferment best at about 12 °C, while red wines need to be at 20 °C or over. Fermentation slows down and eventually stops as the concentration of alcohol reaches 12–13%. Sediment (called 'lees') collects at the bottom of the vat along with crystals of **tartrate**. The wine is decanted into large vats or tanks that have been treated with sulphur dioxide, and is left to settle again. Fresh, sterilised grape juice may be added at this stage to sweeten ('round-off') the wine.

The wine is then left for up to nine months before being bottled. Ageing continues in the bottle with the great Burgundy and Bordeaux wines taking between four and eight years to develop their full potential. After about ten years most wines begin to lose some of their quality, but really great wines may endure for 20 years or more.

Wine consists of hundreds of chemicals apart from water and alcohol. There are other alcohols, **aldehydes**, **acids** and **esters**. There are phenolic compounds such as tannins, amino acids, minerals and aromatic substances. Sparkling wines contain carbon dioxide gas under pressure, and this is generated in a second fermentation by adding sugar and a pure yeast culture. These bottles are kept clamped tight and left for three to five years while pressure builds up, and this may eventually exceed four times atmospheric pressure. The residual yeast is removed by freezing the neck of the inclined bottle and removing the sediment as an ice plug. This is a time-consuming procedure, and a more general way is to carry out the second fermentation in steel pressure tanks and bottle the wine under pressure.

Beer

All that was needed to discover wine was to crush the grapes to release the juice, and the wild yeast spores on the skins would do the fermenting. On the other hand brewing beer is such a complex process that John Postgate in his book *Microbes and Man* says that it is a wonder that it was ever discovered. Yet the first written accounts of beer date from the Egypt of the 1st Dynasty, around 3000 BC.

There is evidence that beer was first brewed in Babylon about a thousand years earlier than this. The clue was traces of the chemical **calcium oxalate** on the inside of a jar. Archaeologists Rudolph Michel and Patrick McGovern, of the University of Pennsylvania, and Virginia Badler, of the University of Toronto, came across the jar while excavating a site at Goden Tepe in Iran. Calcium oxalate is the main component of 'beer-stone' which precipitates from beers in storage tanks. The only other foods rich in oxalate are rhubarb and spinach, neither of which are likely to have been the source of this calcium oxalate.

In fact what the ancient Babylonians were brewing was ale, not beer. Beer requires hops to be added to make it bitter.[5] It is rather remarkable that ale was first brewed 6000 years ago, when we understand the chemistry involved. Ale is produced from fermented barley, yet the starch in cereal grains cannot be fermented by yeasts, which need simple sugars to work on. The first step in brewing ale is to steep the barley in water for a day or two and then leave it in a warm, damp room for a few more days so that it will germinate. This process is called malting. The grain sprouts develop enzymes which break down the starch into simple sugars. The malted barley is next heated to stop the sprouting and steeped in water to soak out the sugars and other plant chemicals on which the yeast can grow. The resulting solution is sieved and is the malt 'wort'.

To make beer the wort is then boiled with hops, which imparts flavour and releases preservative chemicals that prevent the growth of bacteria which might spoil the beer. Finally the yeast *Saccharomyces cerevisiae*[6] is put in, and the whole lot is allowed to ferment for a week or so. As the yeast multiplies it produces not only alcohol but also a whole cocktail of organic molecules which give the beer its flavour. When the fermentation is over, the brew is processed to remove sediment, clarified, and packaged for distribution. Beers for export are often pasteurised. The skill of the brewer is in getting it all right. Brewing so-called real ale, in which there is still active yeast, is still regarded as more of an art than a science.

There are hundreds of brands of beer, although only a few types, which divide roughly into top fermented beers, such as the traditional mild, bitter, pale ale and stout, and the bottom fermented beers, such as pilsner and lager. Ale is made by fermenting malted barley wort with a top-floating yeast which cannot ferment all the malt sugars and which works best above 10 °C. Real ale is then allowed to settle and mature in the cask with the yeast still alive. Extra hops may be added. Real ale does not travel well and has a short life-span. Keg ale is filtered, pasteurised to kill the yeast, and pressurised with carbon

[5] The name beer comes from the Old English word 'beor', which was used in earlier times to indicate an inferior sort of mead.

[6] Most brewers of beer use this yeast, of which there are over 300 strains.

dioxide. It stores well and travels well. Stout is a very dark, bitter, top-fermented beer made from concentrated boiled wort. Porter is a sweet stout, popular with earlier generations but now out of favour and rarely brewed.

Lager is made from less strongly malted barley and is fermented with a yeast (*Saccharomyces carlsbergensis*) which works at the bottom of the wort, and which can ferment all the malt sugars. This yeast will continue to ferment down to 0 °C. Lagers also store well. Fewer hops are added so it is less bitter than beer. Real lager is that produced according to the purity law of Duke Wilhelm IV of Bavaria which was enacted in 1516. This stipulated that only malt, water, hops and yeast could be used. So-called 'lite' beers and lagers have been allowed to ferment until all the sugars are turned to alcohol. Their name is misleading because they are only slightly less calorific than normal beers and lagers.

In chemical terms fermentation is a successful method of synthesising ethanol, and could be the basis of a renewable fuel economy as we shall see. Because fermentation is so successful it has been virtually unchanged since it was first discovered. But that does not mean it can not be made even more efficient, and it may change if genetic engineering can produce a better yeast. Imagine a yeast that could ferment starch directly and tolerate higher concentrations of alcohol before it dies (say 20%).[7] It might even be designed to multiply faster than, or even kill, invading organisms. A yeast that could do all this, and brew a beer that tasted good, would finally turn brewing from an art to a science. Some progress has been made towards this perfect yeast. The world's largest supplier of starter yeasts, the Canadian company L'Allemand of Montreal, has genetically engineered a yeast which has the ability to produce a toxin that can kill rogue yeasts from the wild.

Spirits

To make drinks with a high level of alcohol it is necessary to use a still. Distillation was discovered by the ancient Romans over 2000 years

---

[7] Some yeasts that home-brewers can buy are reputed to produce a concentration of 20% alcohol after about three weeks fermentation of sugar.

ago, and brought to a fine art by the monks of the Middle Ages. Distilling wine drives off the alcohol, and when the vapours, the 'spirits', are cooled and collected you have a product which is mainly alcohol. In fact **methanol** distills off first because this boils at 65 °C, and should be discarded, then the alcohol comes off, boiling at 78 °C. The vapour that condenses from a still also contains a lot of water, but a second and maybe a third distillation will reduce this until we collect almost pure alcohol. Distillation alone cannot remove the last traces of water because water and alcohol form a so-called azeotropic mixture, which means they distill over together once a certain composition is reached. This composition is about 95% ethanol, 5% water. To get absolutely pure alcohol, ways other than distillation are needed to remove the water, but such alcohol is never required for making drinks.

Distillation produces a liquor that burns the mouth and throat, but diluted with pure water to around 40% alcohol it becomes drinkable. A litre bottle of spirits of normal strength consists of 400 ml of alcohol and 600 ml of distilled water, plus traces of essential extras. Many drinks are based on spirits that have been mixed with other things, and the Box opposite lists the contents of some of the better known ones.

In theory spirits are an alcoholic beverage that will do all that beers and wines will do, but freed of impurities. Spirits also act as their own preservative since nothing can live in them, and so no micro-organisms can oxidise their alcohol to acetic acid.

Through the centuries distilled alcohol has been a part of the medical pharmacopeia. The first distilled alcohol was known as *aqua vita* or water of life, and was valued for its restorative properties. Spirits were also used by surgeons as an early form of anaesthetic and to clean wounds. Alcohol was used to wipe the skin before injections because it is a good antiseptic when pure. For the same reason it is also a good preservative.

What distinguishes the different kinds of spirits are the tiny amounts of other volatiles which distill over with the alcohol, and in this way we have brandy (which comes from wine), whisky (which like beer is made from fermented barley), rum (from fermented molasses) and bourbon (from fermented maize). Vodka is made from fermented grain or potatoes, and is the purest drink of all because it is passed through a charcoal filter to remove everything except the alcohol. This is why it is almost tasteless.

Exotic drinks (liqueurs and cordials)

Liqueurs, also known as cordials, have 20–35% alcohol, are sweetened with sugar, glucose or caramel (up to 500 g per litre) and flavoured with various fruit essences, spices and herbs. Manufacturers are under no legal obligation to reveal what they use. Indeed, part of the mystique of speciality drinks is the 'secret recipe', and when most of these were first introduced their secrets were safe. Modern chemical analysis can now reveal exactly what has been used in their concoction. Some are artificially coloured. Here are the main ingredients of the more popular varieties.

Name of drink	Spirit base	Other ingredients apart from sweeteners
Advocaat	gin	egg yolk
Benedictine	Cognac brandy	honey, herbs
Campari	brandy & wine	bitters*
Cherry brandy	brandy	cherry essence
Cointreau	brandy	orange essence
Crème de Menthe	brandy	mint extract
Drambuie	whisky	honey
Malibu	rum	coconut essence
Martini	wine & spirit	mixed herbs extract
Ouzo	raw spirit**	aniseed
Pernod	raw spirit**	aniseed and herb extract
Pimm's No. 1	gin & wine	herb extract
Southern Comfort	bourbon	peach essence
Tia Maria	rum	coffee extract

*mainly extract of wormwood
**can be any distilled alcohol produced by fermentation

Gin is made from fermented grain alcohol, and is flavoured by a second and even third distillation from a blend of herbs, fruits and berries. These so-called botanicals are a secret blend of up to ten fruits, herbs and spices which must include juniper, this being the essential component which identifies a drink as gin. Other botanicals may be coriander, angelica, orris root, almonds, cassia bark, liquorice, orange and lemon peel, cardamom, cinnamon, and nutmeg. After the botanicals have been steeped in the spirit it is distilled a final time. (In some distilleries the spirit is distilled through the botanicals suspended above it.) The first fraction which distills over is discarded. The middle fraction of distillate is collected, diluted to between 47% and 37%, and bottled and labelled as London gin, a term that is protected by international regulations.[8] Distillers guard the blend of their botanicals, although there is really nothing to be secret about, because chemical analysis of the product will reveal all.

Scotch whisky producers on the other hand are very open about how they make their product, safe in the knowledge that all attempts to produce it synthetically have failed, despite the fact that chemists have identified the hundreds of molecules that are present. The flavours of Scotch are introduced at all stages in its production, and even include **phenols** from the peat smoke used to dry the malt. Though most malt is now dried by gas heating, a little peat smoke is always added. Some malt is still dried over peat fires, especially for the whisky that is produced on the island of Islay. The mashing stage which extracts the sugars from the malted barley also extracts other flavours, and the fermentation with yeast produces yet more volatile substances. Many of these are related to alcohol itself, such as methanol, **propanol** and **butanol**, but only in small amounts. None of these adds to the flavour, but they convert easily to aldehydes and these do have flavour, and then to carboxylic acids, which are even stronger and sharper tasting. Most flavour of all comes from esters, which are often fruity to taste when pure, and they are produced by the chemical reaction of alcohols and acids. (In some cases just one substance may provide most of the taste, such as **2-ethyl-3-methylbutanoic acid**, which is what gives rum its characteristic flavour, but for most spirits there are several molecules which provide the aroma and flavour.)

---

[8] Gin can be made simply by adding flavours and essences to distilled spirits, thereby reducing the cost of a final distillation. Such gin cannot be called London gin. ˙

The final stage in the life of a whisky is its stay in a bonded warehouse for at least three years, although this may be up to 12 years or more. Ageing not only leaches out more chemicals from the wood of the cask, but is an expensive process since 2% of the whisky evaporates through the walls of the cask each year. The leached chemicals are aldehydes or acids that are derived from benzene compounds such as the aldehyde vanilla, but there are also wood sugars, **steroids**, tannins and even minerals. All are there in minute traces, but all add to the whisky's complex taste, and make it impossible to forge.

According to Dr John Piggott, senior lecturer in the Department of Bioscience and Biotechnology at the University of Strathclyde in Scotland, the flavour of Scotch is a mystery: 'If you blend all the identified flavour components of whisky together it doesn't make whisky. The secret still eludes us.' Most Scotch is a blend of whiskies, from different distilleries. Even the expensive, so-called single-malt whiskies are generally blends, but of whiskies from different casks in the same distillery. There are some single-cask whiskies, but these are only for the dedicated connoisseurs.

Fraudulent drink

Chemical analysis shows there to be some rather surprising natural chemicals in drink, such as dimethyl sulphide, the chemical responsible for bad breath, and hydrogen cyanide (also known as Prussic acid or hydrocyanic acid). This almond-flavoured chemical is there only in a few parts per million, and too little to affect us. Another component in much larger amounts is methanol, and this can be worrying. It is possible to end up literally blind drunk on illicit spirits with high levels of methanol. Our body turns this into **formaldehyde** and **formic acid**, which attack the retina of the eye. The low level of methanol in wines and spirits is quite harmless.

Chemists are now able to isolate and identify almost all the chemicals in a drink, some of which are present only as parts per billion. In theory with this knowledge it should be possible to combine these chemicals to make synthetic versions, and this is constantly being tried by drinks forgers. Some drinks are easier to forge than others, but it has also become correspondingly easier for analytical chemists to prove that a drink is not what it appears.

While it is acceptable to produce alcohol for industry from agricultural surpluses such as grain and molasses, it is not acceptable to turn the alcohol from industrial plants into drinks. To prevent this happening such alcohol can be 'denatured', and this is done by adding a small amount of a bitter-tasting agents such as sucrose octa-acetate or **Bitrex**, which makes them too unpleasant to drink.[9]

There is no chemical difference between alcohol produced naturally and that made industrially. Fraudsters have been tempted to turn industrial alcohol into wines and spirits, or to boost the level of alcohol in other drinks by adding the industrial variety. Such rogue liquors are easily detected. The tell-tale marker which traps them is radioactivity, or rather the lack of it. The carbon dioxide of the atmosphere which is picked up by grape and grain is composed of three varieties (called isotopes) of carbon. These differ in the number of particles in the nucleus of the atom. The most abundant is carbon-12, which has six protons and six neutrons, and constitutes 98.90% of carbon, while carbon-13 with an extra neutron constitutes 1.10%. There is also present the naturally occurring carbon-14 with two extra neutrons, of which there is a minute trace. Carbon 12 and 13 are stable atoms, but carbon-14 is radioactive and decays with a half-life (see pp 136–137) of 5730 years.

Fossil fuels, which have lain underground for millions of years, have no radioactive carbon-14 left. Consequently alcohol made industrially from ethylene, which is produced from oil, can be distinguished from that made by fermenting carbohydrates. If forensic chemists suspect a drink has been made or doctored with industrial material they simply measure the radioactivity of the alcohol. Too low and it must be phoney.

The non-radioactive isotopes, carbon-12 and carbon-13, can also help in more subtle forms of analysis, together with the hydrogen isotopes, hydrogen-1 and hydrogen-2.[10] Normal hydrogen is

[9] Industrial alcohol used to be contaminated with high levels of methanol (hence its common name of 'meths'). The level of methanol is not such as to constitute a health hazard, and there are down-and-outs prepared to drink meths as a cheap substitute for gin.

[10] Hydrogen-2 is also known as deuterium.

99.985% hydrogen-1 and 0.015% hydrogen-2. Together these isotopes can reveal whether a drink has been doctored in any way at all. The technique, developed at the University of Nantes in France, is so sophisticated that it can identify the country and region of a wine. The method analyses the proportion of the different isotopes of hydrogen and carbon in a sample, and this can be done extremely accurately by a technique called mass spectrometry. The ratio of the isotopes in any sample of alcohol varies according to the type of vine and the environment in which it grows.

The alcohol produced from fermented grape sugars is different from that which comes from the sugars of malted barley, or from fruit such as apples. Using this technique, the researchers were able to show that 17 out of 23 samples of Holland gin contained alcohol from fermented beet sugar instead of from grain. When they did a survey of Beaujolais wines they discovered that some had up to half their alcohol derived from added sugar. The isotope technique can even tell whether the carbon dioxide in a sparkling wine is 'natural', in other words comes entirely from the fermentation, or whether industrial carbon dioxide has been used to supplement the natural gas in order to perk up the wine.

Other methods of chemical analysis can also reveal suspect drinks by identifying compounds which should *not* be present. The spirit which often turns out to be less than it seems is brandy. Every year hundreds of suspect bottles are tested in Germany and almost all turn out not to have been made by the traditional method of distilling wine. Many contain evidence of being made from rotting fruit, a fact that chemists can easily prove by showing the presence of **allyl alcohol** or **acrolein** which are only produced by decaying fruit. Some 'brandy' is pepped up with added esters taken off the shelf of chemical stores, such as **ethyl heptanoate**. A lot of 'brandy' is diluted with cheap wine.[11]

Other cheats try to forge the more sophisticated drinks such as whisky and gin. Exactly what these contain is known, but it is not necessary to add all the components found in the true spirits to produce a passable imitation of the aroma and taste. However, it is the absence of some of the components that is convincing proof that the drinks are phoney.

[11] Cognac is the brandy to choose if you want the real thing, and it must have been made in France and only from wines of the Charente.

It is possible to make wine by blending fruit flavourings, artificial colours and alcohol, but things can sometimes go wrong as they did in Italy in 1986, when 19 people died from drinking cheap wine made with methanol. In Bombay, India, in 1992 during New Year celebrations 85 people died through drinking cheap bootleg liquor, again made from methanol. In the Austrian wine scandal of 1986 the fraudsters did not try to make artificial wine, but to up-grade the quality of low-grade wine by adding the anti-freeze **diethylene glycol**. This had the effect of making the wine sweeter and smoother to the palate. Happily so little of this chemical was needed to boost the quality that no-one died.

The risks and benefits of drinking alcohol

The health risks that a heavy drinker faces are broken bones, obesity, addiction, ulcers, cirrhosis of the liver, brain damage, a particularly nasty form of heart disease called cardiomyopathy, and possibly even cancer of the oesophagus. A more immediate risk may be death, if he or she drives while drunk—and the death may not necessarily be only their own. The upper legal limit for regarding someone as drunk was once based on an alcohol level of 0.15%, equivalent to 150 mg in 100 ml of blood. While this may not incapacitate a person for some activities, such as playing darts, it is far too high for anyone who is driving a vehicle. There are good reasons for believing that even 50 mg is too high a level for safe driving, although some countries permit up to 80 mg.

Alcohol can do serious harm to the human body. The effects on the liver are complex and still not clearly understood. They stem from changes in the levels of co-factors, which are chemicals that are necessary to the action of enzymes. The outcome is that the oxidation of fatty acids is impaired in the mitochondria, the structures in the cells that produce energy, and more fats are synthesised. These fats accumulate in the liver to produce what is known as fatty liver, a symptom of alcoholism. Cirrhosis of the liver is a condition exacerbated by long term abuse of alcohol. It was once thought to be caused by it, but this seems unlikely. Cirrhosis was seen as a toxic response in which normal tissue is replaced by collagen (fibrous tissue). As a consequence the liver functions less efficiently.

Alcoholism is believed to affect about 20 people in 10 000. This condition may be partly inherited according to studies by Ernest Noble of the University of California, Los Angeles, and Kenneth Blum of the University of Texas. They discovered that a certain receptor in the brain was much more common among alcoholics than in the general population. This receptor responds to dopamine, the chemical messenger in the brain that triggers pleasurable feelings. Women are also more likely to develop alcoholism than men, and a study on 1033 pairs of identical female twins has provided further proof of a genetic factor. Kenneth Kendler at Virginia Commonwealth University in Richmond reported in the *Journal of the American Medical Association* that the numbers of twins both becoming alcoholics were much higher than expected, which led him to suspect that hereditary factors may account for this condition in up to 60% of cases.

Alcohol can be regarded as a form of food. A unit of alcohol (10 grams) provides us with about 70 Calories of energy. The recommended maximum intake for a man is 21 units of alcohol a week, and this will provide about 1400 Calories of his energy intake. For a woman the recommended amount is 14 units of alcohol, which will provide around 1000 Calories. While these are not excessive amounts they nevertheless must be taken into account when planning a diet. Alcohol accounts for about 10% of the food intake of adults in the USA and Europe. It was once considered such a convenient way of providing energy that it used to be added to intravenous drips as a supplement.

As a food, alcohol has some disadvantages. It cannot supply energy in the same way as carbohydrates, so it does little to help us work or exercise, but it can supply excess Calories that lead to weight gain, and can be a very expensive form of food. A pint of beer will provide between 140 and 210 Calories depending on whether it is mild, bitter, keg, lager or stout. Draught mild has the fewest Calories (140) while stouts like Guinness have the most (210). The Calorie content of wines and ciders depends also on the sugars they contain, and that can be judged on whether they are labelled sweet, medium, or dry. Four fluid ounces of wine (a normal glass) will be about 110

Calories if it is sweet, but 75 if it is dry. It makes no difference whether the wine is red or white. A glass of sherry is about 50 ml and provides between 55 (dry) and 65 (sweet) Calories. A single measure of spirits or liqueurs is about 25 ml (a sixth of a gill) and provides between 50 and 75 Calories. Whisky and gin are about 50, while liqueurs like cherry brandy and Cointreau are nearer 75.

Alcoholic drinks like beer and wine also provide other dietary nutrients such as minerals and vitamins, but spirits and other distilled drinks lack these. A pint of beer contains about 50 mg of calcium and 0.1 mg of iron, while a glass of wine has only about 10 mg of calcium but up to 1 mg of iron. Cider has almost 3 mg of iron per pint. Beer is also a source of the B vitamins, although wines have less of them. These vitamins come from the yeast. No alcoholic drink provides vitamins A, C or D.

The benefits of alcohol are varied. Some are well known, such as a nightcap of a glass of beer, or a shot of whisky, to help us get off to sleep. However, it may not be a good night's sleep, because alcohol appears to deprive us of a key type of sleep, the early sleep in which we dream. Continued loss of this essential sleep may be the reason why very heavy drinkers eventually suffer the hallucinations traditionally known as the 'DTs' (*delirium tremens*), although these are more likely to strike as part of the symptoms of alcohol withdrawal.

Moderate drinkers, those who take two or three units of alcohol a day, suffer less coronary heart disease and have lower cholesterol levels. Drink may be beneficial for the heart but only in moderation. Men who drink more than the equivalent of six pints of beer a day (12 units) are twice as likely to suffer a sudden heart attack as non-drinkers, according to Professor Gerry Shaper of the Royal Free Hospital in London who carried out an eight-year study of over 7500 middle-aged men. Each year about 25 died this way, and two thirds of these were heavy drinkers.

Studies of the life-style of French farmers supports drinking as beneficial, as revealed in the leading medical journal the *Lancet* in 1992. Dr Serge Renaud of the National Institute of Health and Medical Research in Lyons, France, believes that French farmers are less prone to coronary heart disease, despite a diet rich in animal fats, because of their relatively high consumption of alcohol in the form of red wine. Two glasses a day is the recommended amount.

Red wine is not a medicine, and it will not cure or relieve heart problems that are already present. If there is any value in drinking red wine, it lies in its ability to prevent heart disease. This message of the benefits of red wine increased sales by over 44% in the USA when it was announced in 1992. A group at Kaiser Permanent Medical Center in Oakland, California, claimed that drinkers of white wine were also less likely to suffer coronary heart disease. Whether it is the alcohol itself or phenolic compounds in the wine that are responsible is a matter of debate. The antioxidant behaviour of the phenolics may slow down plaque formation from low-density lipoproteins, the 'bad' form of cholesterol, which we will be discussing in the next chapter.

Not only do drinkers appear to have lower cholesterol levels, and have less coronary heart disease, but they also get fewer colds. That was the conclusion of the final research project carried out by the now-defunct Common Cold Unit at Salisbury, England. Not every common cold virus we catch leads to a cold. Between four and ten viruses infect us every year, but only half of these end up as a bad cold—for the rest we either have existing antibodies against the virus, or we can shake it off easily. If we are under stress this may reduce our immunity. Nor can you stop a cold with the popular cure of hot whisky and lemon—all this does is to make the misery a bit more bearable. However it works its magic, alcohol certainly seems to reward many of its moderate users: those who drink two units of alcohol a day catch half the number of colds that non-drinkers suffer. This does not apply to smokers though; smoking cancels out the protection against colds which comes with drinking.

The medicinal benefits of wine drinking have a long history, and in his quirky books *Wine is the Best Medicine* and *Your Good Health!* the late Dr Eduard Maury set out a careful list of the right wines to treat various ailments from indigestion to arthritis. He advocated a bottle of wine a day (700 ml) for the average 70 kg person, which is about twice the number of units recommended for a male. Some of the supposed benefits of wine are already popular folklore, for example Dr Maury recommends champagne for indigestion, high blood pressure, and as a tonic during convalescence. Less well known are his rather whimsical recommendations of drinking Vouvray for constipation,

Médoc for diarrhoea and arthritis; and Bordeaux or Riesling for obesity.

Low-alcohol drinks

If you are not convinced that there are any benefits to be gained by drinking alcohol, or if you are driving, then you can choose a non-alcoholic wine or lager. Low-alcohol beer can be made simply by stopping the fermentation of the wort when the level of alcohol is still under 1.2%, which is the legal limit for a beer to be so classified. Another way is to ferment the wort with a yeast that cannot digest maltose. Wine yeasts would never normally encounter maltose, as there is none in wine must, and so adding such a yeast to wort will result in an alcoholic content of around 0.5%.

For a beer to be called alcohol-free it must have less than 50 ppm of alcohol (0.05%). This can only be achieved by removing as much alcohol as possible after normal fermentation. Distilling wine or beer is the way to make spirits, and it is also the way to lower the alcohol content of the wine or beer. Distilling beer and wine also changes their flavour, so to prevent this the alcohol is distilled off with much gentler heating by lowering the pressure. In this way the level of alcohol can be reduced to well below 1%. Distillation also removes other components of the beer, and these have to be put back later.[12] The other components which can be lost during removal of the alcohol are the flavoursome esters which the yeasts produce, and again these can be added later.

Removing alcohol may seem a rather thankless task for the producers of beer and wines, since the customer is reluctant to pay the full price for a drink that lacks the key ingredient and the one which they wrongly assume makes drink so expensive to produce. The sophisticated process to remove alcohol explains why such drinks are no cheaper than their normal counterparts. It is a myth that alcohol is what makes drinks expensive. Alcohol is cheap and easy to make, as home-brewers and wine-makers know. It is only expensive to buy in countries where it is heavily taxed.

---

[12] In fact to have more quality control over their products, brewers are increasingly using extracts of malt for colouring beer, and extracts of hops for flavouring them.

Industrial alcohol (ethanol)

A conflict of interests surrounds alcohol (ethanol). Some people regard it as essentially an agricultural product and deserving of special consideration, while others produce it for use as a chemical and they see it merely as an industrial chemical like any other. This conflict leads to political questions of priority and pricing, but one day there may only be agricultural alcohol available. Alcohol production from fermented crops is a natural renewable resource, and it gets our approval for being environmentally friendly, but the real cost is higher than when it is made synthetically in an industrial plant. But whichever way alcohol is produced, chemically it is exactly the same molecule.

To make ethanol industrially, by the reaction of water and ethylene at 300 °C requires a pressure of about 60 times atmospheric pressure, with the presence of phosphoric acid as a catalyst. The rate of conversion of ethylene to ethanol is low, about 5%, and the ethanol has to be separated out and the reaction mixture put through the process again and again.

Alcohol was once a feedstock for the chemicals industry in its own right, and was used to make acetaldehyde, from which acetic acid and acetic anhydride were then manufactured. These were used to make paints, fibres (like rayon and acetate), plastic film (like cellophane), and aspirin. The acetic anhydride to make these products can now be made more efficiently in other ways than starting with ethanol. Alcohol is still used as an industrial solvent, and of course as the solvent for perfumes (see Chapter 1). Alcohol is also used to manufacture various flavours, detergents and disinfectants.

The Brazilians looked to ethanol as their way to overcome the oil crisis of 1973. Alcohol makes a useful liquid fuel, and although weight for weight it does not release as much energy as petrol when it burns, it still releases a lot.[13] In Brazil there are over one million cars which run on neat ethanol, and another million which use an ethanol/petrol mixture in the ratio of 20%/80%. For a time in the 1980s it looked as though all cars produced in Brazil would eventually run only on alcohol, but the price of crude oil on the world markets fell, and

---

[13] A gram of ethanol releases 7 Calories, and a gram of petrol releases 9 Calories—exactly the same amounts of energy as alcohol and fats release in our diet.

once again petrol became the more economic fuel. During the 1980s, home-produced ethanol saved the Brazilian economy over $20 billion, and generated employment for over a million people.

The Brazilians began to turn their surplus molasses into ethanol as long ago as the 1920s, and it was added to petrol to the extent of 5%. At the time many other countries did the same, since it was thought to produce a cleaner burning petrol, although in those days this was seen as primarily of benefit to the engine rather than to the environment. Brazil had a supply of molasses that could be turned into ethanol very easily, and molasses was a surplus by-product of their sugar industry. By 1987 they were producing 25 million tonnes of cane sugar a year, of which 7 million tonnes were used as food and the rest was fermented to make almost 12 billion litres of ethanol. This replaced over a quarter of the petrol from imported oil.

Some remarkable advances were made in sugarcane production. Year by year more and more land was turned over to growing this crop until it exceeded 4 million hectares, although this was only 8% of their cultivated land. In addition crop yields went up almost 20% from 65 tonnes per hectare to 77 tonnes per hectare, and the amount of alcohol produced from this increased by nearly 40% from 2700 litres per hectare to 3700 litres per hectare. This improved productivity came about through better industrial methods of converting the sugar of the sugarcane to alcohol, and by better distillation techniques.

Not only was Brazil producing over 12 billion litres of ethanol by 1988, it was also gaining a modest amount of electricity by burning the sugarcane residues, called bagasse. This was used mainly as heating fuel in the ethanol refineries, but an excess was going to generate electricity. One very noticeable change of all this activity was to the quality of the air of São Paulo (15 million inhabitants), where most cars run on ethanol. This city was no longer afflicted with the kind of smog that afflicts Mexico City (17 million inhabitants) where cars burn only petrol, and where smog and ozone levels pollute the air due to the effects of sunlight on unburnt hydrocarbons from the exhausts of cars.

Another unexpected benefit of the ethanol economy was that the sugarcane needs relatively little input in terms of nitrogen fertilizers— about 80 kg per hectare—although we will discover other unsuspected sources of nitrogen in Chapter 8. You might think that the

Brazilian experiment is an ecological success story, and you would be right. You might applaud it as an environmental success story too, but it is no longer an *economic* success story. That was true only when the price of oil was high. When it fell below $20 a barrel in the late 1980s the Brazilian experiment began to falter, and whereas in the mid-1980s they had talked of producing only ethanol-burning cars in Brazil by the end of the century, by 1993 they were talking of returning once again to the manufacture of petrol-burning cars.

The Brazilian chemical industry was also geared to using ethanol as a feedstock and turning it to ethylene (this is the reverse of the chemical reaction that is more generally used to turn ethylene into ethanol). Other chemicals that are produced from ethanol include acetaldehyde, acetic acid, diethyl ether, ethyl acetate, butanol, **octanol** and **ethyl chloride**. Relatively little ethanol went into the manufacture of these, about 400 million litres a year, but this list of chemicals shows what can be produced from a renewable resource if the need arises.

Sugarcane is not the only carbohydrate-producing plant from which ethanol for industry is currently being made, and in the USA over 3.5 billion litres of ethanol a year are produced by fermenting starch of surplus corn. This looks set to rise to over 7 billion litres by the year 2000, and much of it will end up in reformulated blends of gasoline for use in cities where pollution is a problem.

In a future world which has used up its accessible fossil reserves, we might be dependent on carbohydrates for all our energy and chemicals. Our lifestyle by then might be so closely tied in with the production of alcohol from crops that drinking it may seem merely an antisocial waste of a key natural resource. But for the time being: cheers!

CHAPTER FOUR

Cholesterol, Fats and Fibre

Cholesterol; oils and fats; fibre; hidden fibre (resistant starch); the chemistry of successful dieting

Most people eventually die of heart disease, and many of them are carried off in the prime of life. The causes of coronary heart disease still remain unclear and a bewildering number of suggestions have been advanced to explain it, many based on epidemiological studies of the health records of large numbers of people. A list of over two hundred factors that have been implicated in heart disease was published by P. N. Hopkins and R. R. Williams in the journal *Atherosclerosis* in 1981. Among the non-dietary contributory factors were: being a man; living in a city; snoring; baldness; lack of exercise; an aggressive personality; being rich; being poor; stress at work; and being Jewish.

We have little control over most of these, and so perhaps it is not surprising that attention has focused on dietary contributory factors that have been suggested as the causes of heart disease. Among those which Hopkins and Williams list were some that were thought to be dangerous in excess—such as sugar, coffee, saturated fats, chlorinated water, alcohol and milk—and some that were thought to put a person at risk if they did not get enough, including zinc, fish oils, garlic, unsaturated fats, alcohol (again) and milk (again). Clearly none of these is the primary cause of heart disease, but many of them might be indirectly linked to it. Plainly some of them are way off target, even though they have the support of epidemiological studies.

Protecting the most important organ of our body throughout our life has become the concern of many, and they have been encouraged to believe that this can be done by eating healthy foods, and especially those low in cholesterol, low in animals fats and high in fibre. In this chapter we will look a little closer at the chemistry of these foods, which are given a high profile by food advisers and advertisers, and discover how realistic the claims are.

Cholesterol[1] is now regarded by many people as the number one enemy in the fight against coronary heart disease, and they believe that if they can reduce their cholesterol intake they can reduce the risk of dying this way. Enemy number two is animal fats, and there has been a long campaign to persuade people that these should be avoided and replaced by vegetable oils, which are said to be much healthier on the grounds that they are higher in polyunsaturates and monounsaturates, and lower in saturates.[2] Current food lore is that in addition to avoiding cholesterol and animal fats, we should be taking in more of the great cure-all, fibre. Eat enough of this each day and we can protect ourselves against heart disease. The advice here is to switch from foods such as white bread and rice to their brown counterparts which still retain their natural fibre.

The overall message is that you can protect your heart by knowing about cholesterol, fats and fibre. These three are in fact closely linked: fats in the diet increase the amount of cholesterol in the blood stream, whereas fibre, and especially soluble fibre, will reduce it. However, the *science* behind this simple message is not quite so simple, and the next few pages may reveal a few unexpected truths. Most of the cholesterol in our blood stream is made by the body itself; lard contains more unsaturated fats than saturated fats; and white bread and rice have a lot more fibre than hitherto suspected.

Cholesterol

We all have quite a lot of cholesterol in our blood, and it is there for the excellent reason that it is an essential chemical for the efficient

---

[1] A reminder that when the name of a substance appears in **bold** type, it means that there is more about its chemistry in the Appendix.

[2] Saturated, unsaturated and polyunsaturated are explained later in the chapter.

running of the human body. Only a small amount of this cholesterol comes directly from the food we eat: most of it is made by our own body. Nevertheless, it is not a good thing to have too much. Unfortunately, some individuals have very high cholesterol levels, and the cause is hereditary; about 25 people in 10 000 carry this trait. This is a worrying condition which requires constant monitoring and medical attention to correct it. For such people the battle against cholesterol is never-ending because they are prone to heart disease. The rest of us are lucky by comparison, but that is little comfort to many people who now fear that they too are at risk unless they can reduce the amount of cholesterol in their blood. If you find your blood cholesterol level is too high you can generally reduce it by following the advice of a trained dietitian, which generally means eating less fat and more fibre, and especially soluble fibre. Alcohol may also help reduce cholesterol as we saw in Chapter 3, although this is more likely to be a self-prescribed treatment.

Cholesterol is not a life-threatening toxin, but a medium-sized molecule that is really a building block for important parts of the body. In particular it is an essential component of cell membranes. Cholesterol also stabilises a cell against temperature changes. It is a major part of the membranes of the nervous system, the brain, the spinal cord and the peripheral nerves. In particular it is incorporated into the myelin sheath that insulates the nerves from the surrounding tissue. Cholesterol is also the forerunner of important hormones such as the female sex hormone, oestradiol, and the male sex hormone, testosterone, and of vitamin D, which we need in order to utilise calcium and form bone. Nearly all body tissues are capable of making cholesterol, but the liver and intestines make the most. We require cholesterol to produce the bile we need to digest the fats in our food, and the name itself comes from the Greek words for 'bile solids'.

Essential though cholesterol is, there can be too much of it, and too much causes a build-up of deposits in the arteries, constricts them, and may even block them, with dire consequences.[3] The causes which are now seen as contributing to higher-than-normal cholesterol levels are:

---

[3] Worse still, if a piece of this deposit breaks free it could stop the flow of blood to the heart muscle.

hereditary factors, which are the most important; then high blood pressure, stress, smoking, obesity and dietary cholesterol.

We can do nothing about the first of these since it is written in the genes we inherited from our parents. It is also difficult to change the pressures of work, but we can give up smoking and we can change our diet and lose weight, and maybe reduce the level of blood cholesterol at the same time. Dieting as a course of action is relatively easy to undertake and perhaps this is why it has received most attention, although the focus of that attention has tended to be directed at cholesterol. This is perhaps not surprising in view of the known danger of heart disease suffered by those with naturally high levels of cholesterol in their blood. In their case the high blood cholesterol level can be brought down with medical treatment using potent pharmaceutical drugs, but for most people a change of diet may be all that is required. The drug treatment works in that it prolongs the life of the high risk group, and there is a growing tendency to believe that it will do the same for the rest of us. Regular cholesterol monitoring is possible, even using a home test kit. The simple message that most people have now picked up is that cholesterol is bad news and that foods with high levels of this should be avoided.

Only a little of the cholesterol in our blood comes from our food. The eating of oils and fats has much more effect in boosting it. Our bodies produce bile acids in order to digest these food molecules, and bile acids are made from cholesterol. The more fat we eat, the more bile acid we need. The more bile acid we use, the more cholesterol we make, and the more of this gets in our intestines. When it has done its job it can be reabsorbed by the body and reused.

Food manufacturers have responded to the public's heightened awareness of cholesterol as a dangerous chemical, and today you can even see 'no cholesterol' or 'low cholesterol' on some product labels on supermarket shelves. The public has got the message, but is it a helpful one? Cholesterol itself does not cause heart attacks or high blood pressure. We make four times as much cholesterol a day as we could possibly absorb from our food, even assuming we were to eat a lot of dairy foods such as eggs and cream which contain relatively high amounts.

Cholesterol is so important that together our cells, intestines, and especially our liver make about a gram of it every day. The average adult has 150 grams (5 ounces) in their body, enough to fill a wine glass. Very little of this has come from the cholesterol in their food, and if a person is a vegan vegetarian and eats no animal products, they take in none at all. Their own bodies have to produce all the cholesterol they need. Those who eat meat, fish or dairy products generally consume about half a gram of cholesterol a day, and their body can absorb a little of this. How much depends upon other components of their diet, such as soluble fibre which scavenges cholesterol.

In theory the ideal level of cholesterol in the blood is 200 milligrams per 100 millilitres of blood, and this is the target to aim for in the USA. In Europe the magic number is 5.2, with the cholesterol being measured in the chemical units of millimoles per litre of blood. These numbers are just different ways of expressing the same amount.[4] Because cholesterol is chiefly a hydrocarbon-type molecule, it is insoluble in water, yet it has to be transported by the blood. To do this it combines with lipoproteins, tiny globules which are a combination of fatty acids and proteins. The low density lipoproteins (LDLs) are where most of the cholesterol in our blood is to be found, and this explains why LDL is regarded as threatening because it deposits as plaque and clogs up our arteries, and may even trigger the formation of a blood clot. There is also another type of lipoprotein called high density lipoprotein (HDL), but this is considered less harmful—indeed some claim it acts as a protective factor. HDL is thought to be responsible for removing cholesterol and transporting it safely to the liver.

Given a balanced diet, our bodies regulate the production, use and removal of cholesterol, and it stays at around the 200 mg (or 5.2 mmole) level. There are several factors that may push it above this figure, such as a fatty diet, smoking, stress, and drinking too much coffee.[5] There are some foods which can reduce our cholesterol, such as garlic, alcohol, oats and baked beans. These last two contain a lot of soluble fibre, and a diet which includes them regularly will reduce your blood cholesterol level by about 10%.

---

[4] A millimole of cholesterol is 386 milligrams. This weight is based on the chemical formula of cholesterol.

[5] Not all coffee does this, only coffee which has been kept hot for a long time.

People with naturally high levels may have cholesterol in excess of 300 mg (or 8 mmole), and some even exceed 800 mg (20 mmole). The cause is a hereditary defect in those receptors of the liver that remove cholesterol from the blood. In addition to controlling their diet such people need drug therapy, and there are two kinds of drug they can be prescribed. One type binds the cholesterol in our intestines and prevents it from being absorbed. This type is a more effective form of scavenger than soluble fibre, and requires the patient to take an ounce or two of the prescription a day. To compensate for this loss of cholesterol, the body then uses more cholesterol to make bile and so the level in the blood falls. A modified form of wood cellulose has also been discovered to lower cholesterol by up to 33% within two weeks. The material HPMC (short for hydroxypropylmethylcellulose) is manufactured by Dow Chemical as a thickening agent for cheesecake and desserts, but health and safety tests revealed quite by accident that it dramatically lowers cholesterol levels, and especially LDL levels, by as much as 50%.

The other treatment aimed at reducing cholesterol uses drugs and actually interferes with the way cells produce this chemical; in other words the drugs turn off the supply at source. These pharmaceuticals now have worldwide sales in excess of $1 billion annually, and they are one of the fastest growing drugs. The anti-cholesterol drugs most commonly used are **lovastatin** and **mevastatin** which are derived from a natural substance found in fungi. They work by deactivating an enzyme that is needed to produce **mevalonic acid**, the molecule from which cholesterol is made, and it will reduce cholesterol levels by up to 50%.[6]

Only the few people with naturally high levels of cholesterol really require such a drug; the rest of us can keep ourselves at the normal level with very little effort, just by eating less fat and more fibre. A common course of action is to target high cholesterol foods and cut out dairy products such as butter and eggs, although the effect this has on the level of cholesterol in the blood is minimal. In fact four ounces of chicken contains more cholesterol than an ounce of butter. The table below gives the cholesterol level in several foods. Eating those

---

[6] It is slightly toxic towards the liver.

Cholesterol content of some foods

Food	Amount of cholesterol (grams per 100 g)	Food	Amount of cholesterol (grams per 100 g)
Brain (offal)	3.10	Cheddar cheese	0.07
Egg yolk	1.26	Fish	0.05
Kidney	0.70	Bread	0
Caviar	0.50	Vegetables	0
Butter	0.23	Fruit	0
Chicken	0.07	Margarine	0

high in cholesterol should not worry us, because only a small fraction of the cholesterol we eat ends up in our blood stream.

Fat intake is much more important in determining cholesterol levels because it stimulates bile production which in turn stimulates cholesterol production. Not only that, but the fat that is absorbed by the gut is then the raw material for more cholesterol to be made in the liver. Cooks and chefs of the future may well design meals which combine high cholesterol foods with foods high in soluble fibre.[7]

Reducing cholesterol in the diet is not without its risks, however, as an epidemiological survey in 1992 discovered. Fifty thousand Swedish men and women over 45 who lived in Värmland were studied over a period of 20 years by Dr Gunnar Lindber of the Centre for Public Health Research at Karlsttad. The results were not quite as reassuring as people had hoped: the group of men with low blood cholesterol did not have a much lower death rate than the group with high cholesterol levels. Admittedly they were less likely to die of heart disease, but they made up for this by being more likely to commit suicide or to die prematurely in accidents. Curiously these effects were not found in women.

For the rest of us the best advice, if we suspect our cholesterol level is too high and is putting us at risk, is to eat less fat and more fibre. This advice is also the same as that of many weight-reducing diets, and since being overweight is now thought to put a greater strain on the heart than cholesterol, the advice is still good, although for the wrong reasons. At the end of this chapter I will show how a little chemistry

[7] For example caviar is very high in cholesterol, so what better way to deal with this than to serve this delicacy with baked beans? Stranger things have happened!

can enable you to lose weight quickly and effectively, albeit using the world's most boring diet.

Oils and fats

If we have a diet that has a high level of fat then we are more likely to suffer heart disease, but the connection between them is not a simple cause-and-effect relationship. This can be illustrated by comparing heart disease deaths in countries where fats make up about a third of calorie intake. This includes some of the most prosperous, such as Switzerland, where the death rate from heart disease is low at 30 deaths per 10 000 people per year, while in Germany it is 40, Sweden 60, UK 70, and the USA 80. The French also have a high fat diet, yet they have the second lowest death rate from heart failure of only 20 per 10 000. The Japanese are lower with about 10 per 10 000, but they eat much less fat.

Faced with such a spread of results there clearly must be other factors at work. Some nutritionists, backed up by epidemiological evidence, say that it is the *type* of fat which is the key. According to them it is animal, or saturated, fats which are to blame. These are the real villains, whereas vegetable oils, which are unsaturated, are fine. They found statistical support for their ideas and this showed that there was a slight increase in heart disease among those who ate a lot of dairy products. The evidence fell far short of scientific proof, but it convinced those who wanted to be convinced, and gave them an excuse to preach to the rest of us. Their simple message went forth: animal fats are saturated and so are bad, vegetable oils are polyunsaturated and so are good. Perhaps not surprisingly the facts about fat are much more sophisticated than these simplistic slogans, which show a lack of understanding of their chemistry.

Saturated and unsaturated fats

Even if you already know about saturated fats, it may come as a surprise to learn that butter and cream contain a lot of *unsaturated* fat. It is highly misleading to talk as though all animal fat was saturated, and therefore a health risk, whereas vegetable fats are unsaturated, and therefore good for the heart. Indeed some confectionery contains coconut oil which is almost entirely *saturated* fat. Puzzled?

The words saturated, unsaturated and polyunsaturated are used to describe fats, yet many people have no clear idea of what these chemical terms really mean. Although the phrase 'saturated fat' sounds to be more fattening than 'unsaturated fat', these terms have nothing at all to do with the quantity of fat or its calorie content. Our normal idea of something being saturated is of a sponge saturated with water, but this mental picture is misleading when we talk of saturated fat. Saturation is an old-fashioned chemical term, which is related to the bonds between atoms. The chemical explanation of saturated and unsaturated oils and fats is given in the Box below.

What is meant by saturated and unsaturated oils and fats?

From a chemist's point of view, oils and fats belong in the same class of chemical compounds, the **triglycerides**, which are derivatives of **glycerol**. You might know this compound by its older name of glycerine. It is a triple alcohol (it has three alcohol groups), and if a long-chain fatty acid is attached to each of these alcohol groups then we have a triglyceride. This may be a liquid (an oil) or a solid (a fat). The fatty acids consist of a line of carbon atoms, and it is these carbons which are described as saturated or unsaturated.[8]

Single bonds between atoms are called saturated, whereas double bonds are said to be unsaturated. If we have a chain of carbon atoms each carrying two hydrogen atoms then the chain is depicted in the following way: $—CH_2—CH_2—CH_2—CH_2—$ etc. This is called a saturated chain. However, if adjacent carbon atoms are joined by a double bond ($=$), with a hydrogen missing on each carbon, then the chain is said to be unsaturated and is shown thus $—CH_2—CH=CH—CH_2—$ etc. We can saturate such a chain by reacting it with hydrogen gas.

When there is one double bond in the chain we have a mono–unsaturated material—the name comes from the Greek word *monos* meaning

[8] A saturated carbon atom forms four chemical bonds linking it to four surrounding atoms. If it is bonding to only three surrounding atoms (which means a double bond to one of them) then it is described as unsaturated.

single. If there is more than one double bond in the chain we have a poly-unsaturated fat, from the Greek word *polus*, meaning many.

Generally each of the three long-chain fatty acids in a triglyceride is different. In animal fats most of them are saturated and the triglycerides are solids; in vegetable fats most may be unsaturated, and these triglycerides are liquids. Each type of fat has its advantages, and in terms of food storage, solids are better than liquids because they resist attack by oxygen and so do not go rancid. If we want to reduce the risk of this happening, we can add the missing hydrogens to unsaturated fats and produce 'hydrogenated vegetable oils' or 'partially hydrogenated vegetable oils'. The French chemists of the last century discovered margarine this way. There is nothing sinister in doing this—all that is happening is that a triglyceride of vegetable origin with perhaps two of its long-chains unsa-turated is being turned into a triglyceride that has only one chain unsa-turated. It then resembles a fat of animal origin.

No fat is entirely saturated or unsaturated, whether it be of animal origin or not. Certain types of fat contain so-called 'essential fatty acids' which are vital to various body functions and which we cannot manufacture. These are the polyunsaturated fats such as **linoleic acid**. We have to get our supply of these from what we eat, so a totally fat-free diet would be bad for us.

Some of the information in the table below may surprise you. It shows that in general vegetable oils have more unsaturated fats than saturated fats, but not all vegetable oil is polyunsaturated and not all

Saturated and unsaturated components of oils and fats (triglycerides)

Triglyceride	Saturated (%)	Unsaturated (%)		
		Total	Mono-unsaturated	Polyunsaturated
Rapeseed oil	6	94	64	30
Sunflower oil	11	89	20	69
Olive oil	14	86	77	9
Lard	41	59	47	12
Palm-kernel oil	51	49	39	10
Beef dripping	52	48	44	4
Butter fat	66	34	30	4
Coconut oil	92	8	6	2

animal fat is saturated—far from it. It reveals that lard is mainly an *unsaturated* fat while palm-kernel oil is mainly *saturated*—just the opposite of what you might expect. Coconut oil is really in trouble, but how often has the message to avoid animal fats ever warned you against eating sweets and candy made from coconuts?

In proportion to their weight, humans have at least ten times as many fat cells in their bodies as most other mammals. In the developed world the tendency is to eat a diet with a lot of fat, and it makes up about 40% of our food on average. We must have some fat because it is a natural source of the fat-soluble vitamins A, D, E and K, and we also need essential polyunsaturated fatty acids such as linoleic acid for growth and tissue repair. The current trend is to reduce the fat-content of our diets to a more healthy 30%, of which only a third should be saturated. For a normal person, taking in about 2200 Calories of food energy per day, their fat intake should contribute 700 Calories, which is equivalent to 80 grams or three ounces of fat or oil. Most of this will be part of other foods, and many common items contain a lot of fat. The next table lists a few examples of foods that have more fat than people imagine. It also gives their Calorie content, which again might be rather surprising. Few people would guess that half an avocado pear will provide 20 grams of fat, and a snack-size packet of nuts will provide 15 grams. A fried egg has 8 grams, but then a boiled egg has 6 grams anyway. Eating a lot of fat provides Calories we do not need, and this encourages the body to make and store fat of its own.

Foods containing more fat than is generally realised

Food	Portion weight	Fat content (%)	Calories
Cashew nuts (20 nuts)	40 g	45 g	220
Cheddar cheese	40 g	35 g	170
Chevda (e.g. Bombay mix)	100 g	32 g	490
Croissant*	50 g	20 g	180
Candied popcorn	100 g	20 g	480
Digestive biscuit (two)	30 g	20 g	150
Individual fruit pie	110 g	15 g	410
Avocado pear (half)	130 g	25 g	200
Pilau rice	200 g	11 g	420

*A doughnut is about the same

If animal fats are so bad for us it seems strange that mammals evolved to store a large part of their surplus energy in the saturated form. Even so, less than half is of this type. Humans store mainly unsaturated fat: the body fat of an average adult is about 40% saturated, 60% unsaturated, just as in lard. Of the unsaturated fat all but a few per cent is mono-unsaturated. These ratios will vary a little depending on the individual and on their diet, but clearly we store a lot of our fat as saturated fat because there is some advantage in doing so, perhaps because such fats are more stable to the damage that oxygen can cause. Unsaturated fats are susceptible to chemical reaction with oxygen, and this is why cooking oils require antioxidants, whereas cooking fats have a longer shelf-life and keep without the need of protective chemicals. One curious epidemiological observation uncovered in the saturated versus unsaturated controversy was that people who ate more saturated fats had fewer cancers.[9]

So what should we do about fats? The answer is first to cut down all fats if we are overweight, and then to worry about the saturated-to-unsaturated content of what we eat. This change need be no more difficult than simply changing from butter to margarine and from red meat to white fish. It would not hurt in any case to change to a low-fat spread, and that will generally be made from something like sunflower oil. But don't expect these simple changes to do much in the way of keeping you out of a cardiac unit in hospital, or making you live to be a hundred. If you want to lead a longer, healthier life then you need to do more than make a token gesture of changing from 'unhealthy' saturated animal fats to 'healthy' polyunsaturated vegetable oils. There is no substitute for losing weight if that is your real problem.

Fibre

'Today Americans and Europeans are bombarded with high-powered advertisements by health-food advocates, food faddists, food manufacturers, phar-

---

[9] A reason for this might be that the chemical attack of oxygen on the double bond of an unsaturated fat could form a free radical, a type of compound that is known to trigger cancer.

maceutical manufacturers, and lesser-known charlatans, who offer us what we want to hear about the fibre cure-all, at prices we're willing to pay.'

ROBERT ORY, *Grandma Called it Roughage*

Fibre is the indigestible component of our food, and it has undoubted benefits in preventing constipation (more politely called irregularity) and piles (more politely called haemorrhoids). These afflictions are very common and are exacerbated by today's sedentary lifestyle. Fibre is good because it bulks out stools and keeps them soft and moist so they are easier to pass. Even if fibre only made defecating less of a strain it would be worth making sure our diet contained enough of it, but fibre is reputed to be able to prevent the two big 'Cs': cancer, particularly of the colon, and coronary heart disease. The theory is that it prevents cancer of the colon by removing waste products that may harbour carcinogens, and prevents coronary heart disease by removing cholesterol.

Whole books have been written about fibre and fibre-based diets, and as you might expect they carry a lot of anecdotal evidence in support of eating more fibre. If you worry about high blood pressure and heart trouble then you could do worse than taking in more fibre, and soluble fibre at that, in the knowledge that it is completely harmless, and might do you some good. It is a natural cure for constipation and may prevent some of the problems that are attendant upon this condition.

The man who rediscovered fibre

Hippocrates, the Greek physician who lived around 400 BC, recommended wholemeal bread for its beneficial effects on the bowels. His advice was sound because wholemeal flour contains a lot of fibre. The current craze for fibre can be traced to the late Dr Denis Burkitt, who gave his name to a form of cancer, Burkitt's lymphoma, which affects lymphoid tissue, and which is prevalent in tropical countries. It was while he was working in Africa in the 1960s and 1970s that Dr Burkitt noticed that native Africans rarely suffered from cancer of the colon or rectum, and this he attributed to the ease with which food passed through their bodies, which in turn reflected the large amount of fibre in their diet. Indeed, many of the ailments that he had come across while practising in London, such as piles, diverticular disease, hiatus hernia, appendicitis, atherosclerosis, circulatory disorders and even varicose veins were unknown among rural Africans whose diet of coarsely ground grains, beans and peas was high in fibre. Dr Burkitt then put forward his theory that Westerners who wished to avoid these ailments should likewise take in much more fibre with their food.

What goes into our bodies also has to come out again. We never tire of talking about the food we eat, but social etiquette forbids us talking about the other end of the process—expelling the waste products. Happily we come into the world endowed with a superb system of waste disposal. In some ways it is even more sophisticated than the intake system in which everything enters the same channel. All we have to do with food is find it, chew it and swallow it—after that our body chemistry takes over.

Because atoms can neither be created nor destroyed in chemical processes, those we take in as food have to be got rid of somehow. We dispose of them as gases, liquids and solids. Carbon, the main component of our food after water, can be breathed out as odourless carbon dioxide, and every breath removes this to the extent of about 17 litres of the gas per hour. We also get rid of a little sulfur this way if we have eaten onions or garlic, which have a high sulfur level, and then our breath smells unpleasant because of these volatile sulfur compounds. In addition we can expel carbon as methane gas, again accompanied by sulfur compounds, in other unmentionable ways.

Soluble waste materials, along with the surplus water from the food and drink we consume, exits via our urine. We get rid of water and some waste molecules when we sweat, and again some of these waste molecules, such as butyric acid, are a bit malodorous. But it is the excretion of solid waste that causes us the most worry. It can also cause us the greatest pain if it becomes dehydrated so that we retain it too long and have difficulty ejecting it. This leads to haemorrhoids, which are swollen veins around the anus that can bleed and become infected. One way to prevent this is to eat a lot more fibre.

So what is fibre? The simple answer is indigestible material in our food, of which the main chemical type is a form of carbohydrate called **cellulose**. We produce enzymes in our saliva, stomach and pancreas to break down our food into its basic units. We have various enzymes to digest proteins, fats and normal carbohydrates, but they are powerless against fibre. Although most fibre has no food value, it still serves a useful purpose by providing bulk for our intestines to work on. Some, like lignin and cellulose, passes through virtually unchanged, while some may be partly fermented by the bacteria which live in our lower intestine.

Somewhat surprisingly cellulose is made from glucose, the same as sugar and starch, but in this form of fibre the glucose units have been linked together chemically in such a way that the enzymes in our bodies cannot attack. There may be up to ten thousand glucose units in a cellulose chain. Plants turn glucose into cellulose, and from it they construct their roots, stalks and leaves. They also make other indigestible carbohydrates for the shells and husks which protect their seeds. Cows, sheep and goats can digest cellulose, which is why they feed happily on grass, and they do this by having two stomachs. In one of these stomachs live microbes which can digest cellulose by using enzymes to break it down into glucose units. Single stomach animals like pigs, hens and humans have neither the enzymes nor the microbes with those enzymes to enable them to live off cellulose.

Cellulose is not the only carbohydrate we refer to as fibre, although it is the main one. There are other kinds of indigestible saccharides, such as the pentosans and pectins, which are found in soft fruits and natural gums, and lignin, which is a woody material. All these chemicals count as fibre and they serve various purposes within a plant, with some plants producing a lot of one kind. They differ chemically in their behaviour in water, which is why some are used in food. The least soluble and most abundant is cellulose, although this too has a great capacity to attract water. We make use of this property with paper products such as kitchen towels, nappies and handkerchiefs, all of which are virtually pure cellulose. Some forms of fibre are extremely good at binding to water, and these are referred to as soluble fibre.

Some dietary advisors maintain that soluble fibre clings best to bile acids and ensures that they are excreted, so that the cholesterol is not reabsorbed into the blood stream. Bile acids pass from the gall bladder and mix with food to help break up the fat and oil globules so that enzymes can better digest them. When the bile has served its purpose it is reabsorbed by the gut and stored until another fatty meal comes along. When soluble fibre prevents the bile acids being reabsorbed the liver has to use cholesterol to make more, and the amount in the body falls.

Some foods have a lot of soluble fibre, such as oats and natural gums like gum arabic which is widely used and comes from the stems of *Acacia Senegal* and related species of trees which are native to Sudan, West Africa and Nigeria. It consists of a carbohydrate polymer built from **galactose** and glucoronic acid to which are attached sugars such as **arabinose**. Other forms of soluble cellulose come from seaweed (the alginates and carrageenan), and yet others from beans (for example, guar gum). The bacterium *Xanthomonas campestris* produces xanthan gum, which like cellulose is built of glucose chains with other saccharide units attached. Foods which include these gums will lower the blood cholesterol level.

Gums are often used in food to provide the right texture ('mouthfeel'), rather than being there merely as soluble fibre. Cellulose can also be used for this purpose if it is made water soluble, and that can be done by turning it into methylcellulose. Next time you see this word on the side of a packet of cake mix, or a bottle of salad dressing, remember that it may look rather chemical, but what you are about to eat is perfectly safe and it will provide you with a little extra fibre. It has probably been made from wood cellulose which is chosen because it has no flavour of its own.

Fibre in foods

Food	Fibre (%)	Soluble fibre (%)
Wheat bran	44	3
Oat bran	28	14
All-Bran	27	2
Dried apricots	24	5
Baked beans*	15	6
Rolled oats	14	8
Cornflakes	12	7
Raspberries	7	4
Sweet corn	6	2
Apple	2	1
Brown rice	2	1
White rice	1	see the text

*Canned with tomato sauce. Beans by themselves are usually about 25% fibre

Methylcellulose behaves like so-called soluble fibre, the type that has great affinity with water and which when dissolved in water leads to very viscous solutions.

So what foods should we eat to take in more fibre? Plant foods contain fibre; animal products do not. Some foods, such as bran, oats, apricots and beans, have a great deal of fibre. Ideally we should take in 32 grams of fibre a day (just over an ounce). This does not seem much, but most people in the West take less than half this. The table on the previous page lists a few foods that are commonly cited as being high in fibre. Not all are.

Baked beans and porridge, once derided as staple foods of the very poor, are very good ways of taking in fibre and soluble fibre.[10] Today bran-based and oat-based breakfast cereals like All-Bran and muesli are popular, and brown bread and brown rice are much praised by food advisors. Books about fibre-based diets appeared in the 1980s and sold in millions. A lot of their advice about food was basically sound. A lot was written in ignorance.

White versus brown

The general recommendations for a healthier diet could be summed up thus: change your eating habits from white foods to brown. Some people now believe that brown bread, brown rice and brown sugar are much better than the white varieties. A few people even prefer eggs with brown shells, if they still buy them at all. Although they are an excellent food for all ages, eggs are now avoided by many merely because of the cholesterol they contain.

Ironically, our parents' and grandparents' general guide to healthy eating was to avoid colour. White was right. White was a symbol of purity. The 19th century had seen many food scandals, as producers and shopkeepers adulterated food with everything from sawdust to powdered bone. White foods were thought to be the hardest to cheat on, and they were among the most nutritious; milk, sugar, lard, flour, bread, salt and even white shell eggs were thought to be best. Sugar and salt are among the purest of all because these are crystallised from water, a process which chemists use to remove impurities.

---

[10] Baked beans are navy beans, which are grown in the USA.

Brown eggs: Let us begin with the easiest myth to dispel: brown eggs are better than white. It is often popularly imagined that battery hens, which are fed only on cereals, lay eggs with paler yolks and white shells because they lack some vital nutrients which free-range hens get from the seeds, grass, insects and grubs they eat. Diet does affect the eggs that a hen lays, but only marginally, and brown eggs with deep yellow yolks also come from battery hens. We may object to any animal being raised in unnatural conditions merely to provide humans with cheaper food and farmers with higher profits, but that is another issue, and little to do with the chemistry of eggs. The colour of a bird's egg depends upon the species, and so it is with hens. It has no nutritional significance. Brown or white, battery or free-range, eggs are just eggs.

Brown bread: Bread is baked from three types of flour: white, brown and wholemeal. Wholemeal bread has to be made from the whole wheat grain, so that it contains the maximum amount of fibre. In white flour the outer third of the grain is removed as bran, and along with it goes most of the fibre. Brown bread is intermediate between these two extremes, with about 15% of the bran being removed. Some bread can have extra fibre added, and some can be baked with malted flour when it is described as granary bread.

A slice of bread, be it white, brown or wholemeal, provides about the same amount of energy, protein, fat, carbohydrate, iron and B vitamins. They differ in the amounts of calcium and fibre they provide: a slice of white bread has 40 mg of calcium and 1.5 g of fibre; a slice of brown bread has 35 mg of calcium and 2 g of fibre; and a slice of wholemeal bread has 20 mg of calcium and 2.5 g of fibre. The calcium of white bread should make it the preferred choice for children and pregnant women. For the rest of us it is the fibre which may influence our choice. If we aim at 32 g of fibre a day then switching from white to wholemeal bread will go a little way towards this total, depending on how much bread we eat. But even here the superiority of brown over white is partly an illusion, as we shall see when we talk about hidden fibre (see p. 108).

Brown sugar: If you like your warm drinks sweet, then brown sugar gives a richer taste in coffee but spoils the more delicate flavour of tea. If we have to choose between brown or white sugar, then taste alone is our only guide, because in all other respects they are the same. As we saw in Chapter 2, the brown component in sugar is the sort of chemical that forms when we heat sugar and caramelise it. Sometimes sugar and molasses are simply mixed to give the various kinds of brown sugars that are called by a variety of names, such as demerara, which is nice in coffee, or dark Muscovado, which has a stronger taste and is ideal for spicy cakes and gingerbread. All are virtually pure carbohydrate, all provide the same amount of energy per spoonful, and the only difference is the taste which in brown sugar comes from traces of many chemicals, such as complex saccharides, salts, organic acids and amino acids.

Brown rice: Brown rice is to ordinary rice what brown bread is to white: richer in fibre. The brown variety also provides a little more protein, calcium, vitamins and fat, but that was not the reason for encouraging people to make the switch. Brown rice needs cooking for longer and is harder to chew, but some believe this is a price worth paying in the battle for more fibre.[11] Closer inspection shows that it is not really worth the effort. Brown rice certainly has almost twice as much fibre as white rice, but that is still not a lot. Boiled brown rice is 1.5 per cent fibre, while white rice is only 0.8 per cent. A normal portion of brown rice will provide you with 2.5 grams of fibre, and white rice with only 1.3 grams. However, white rice has more fibre than these figures suggest, as we shall now see.

Hidden fibre (resistant starch)

Those who advocate a change to brown rice did not realise that when white rice is cooked something rather wonderful happens: the amount of fibre actually increases. The same is true of products

[11] It is probably worth the struggle if this is your only source of B vitamins, of which brown rice has a lot and white rice very little.

Cooked foods high in resistant starch (hidden fibre)

Food	Resistant starch*
Boiled white rice	78%
Kelloggs' cornflakes	64%
Spaghetti	38%
Boiled potatoes	33%
White bread	29%
Peas	21%

*Resistant starch as a percentage of the total fibre content

made from white flour. Removing the outer layer of the wheat or rice does remove most of the plant fibre. What is not appreciated is that this is compensated for in another way, with hidden fibre or, as it is more correctly called, resistant starch, which forms on cooking. It is called this because it resists digestion by the enzymes of the stomach and intestines and passes out of the body with its energy store intact.

A dish of a supermarket own-brand of cornflakes will provide you with half a gram of dietary fibre. On the other hand, if you choose Kelloggs, the world's best selling proprietary brand, you will get twice as much fibre. Yet both are made from the same type of flour; the only difference is in the way they are cooked. The traditional high temperature method used by Kelloggs converts some of the carbohydrate into resistant starch. This hidden fibre was discovered in 1982 by food chemists at Cambridge, England, and now it has turned up in all sorts of cooked foods such as potatoes, pasta, peas, beans, and especially rice. Scientists believe that it can be considered as a form of dietary fibre, and may even be a superior form, as we shall see. Some foods contain remarkably high levels of resistant starch, and the table above shows those foods in which it accounts for a fifth or more of the fibre they contain. All-Bran for breakfast can provide a third of our daily fibre needs, but so can an evening meal of white rice and peas.

Resistant starch explains some other mysteries of cooked food. It was known in the last century that when some foods were re-heated their nutritional value went down. What was hard to explain was why this happened to mashed potato, but not to carrots. The answer came to light when a team led by Hans Englyst, of the Dunn Centre, Cambridge, was able to show that this effect could be explained by the formation of resistant starch from ordinary starch.

Colin Berry, head of nutrition and food safety at the Flour Milling and Baking Research Association in the UK, has researched resistant starch and looked at ways in which this could be deliberately increased. His work has shown that it needs three things for its formation: there has to be starch present in the food; it needs a high cooking temperature; and the food has to have a high moisture content. When these conditions come together they allow some of the starch molecules to bind so closely that our digestive enzymes cannot attack them. Once resistant starch has formed it cannot be changed back into digestible starch. Indeed, reheating food will only serve to make more of it.

For years, official tests for dietary fibre ignored resistant starch, and so the tables of fibre content tended to be on the low side. Recently the published fibre level in many foods has increased, much to the embarrassment of some dietary advisors, because it has increased most among the foods which they said should be avoided. Colin Berry claims that it is now possible to have foods with high levels of resistant starch that could actually be better for us than the brown counterpart, even though officially they contain no dietary fibre at all. Rats fed a diet rich in resistant starch for the whole of their natural lives were found to have fewer bowel cancers than rats fed conventional high fibre foods. The resistant starch in rice might also explain why the Japanese suffer much less from diseases of the bowel, such as cancer and diverticular disease, than we would predict from their preference for white rice.

The chemistry of successful dieting

Dieting is surrounded by a mystique that needs to be swept away. A little knowledge and you can learn how to do it effectively, effortlessly and cheaply. You will also learn that you can cheat yourself, but not the laws of chemistry.

Dieting is probably the most popular, but least enjoyed, activity in developed societies. It is nevertheless fascinating because it is easy to do, it provides hours of innocent agony and is a good topic of conversation—and yet we are not really expected by our friends to do it successfully. There are hundreds of books devoted to dieting.

These advise us how to lose weight, but they require us to monitor what we eat very carefully, and some seem to regard food almost as if it were medicine, to be measured carefully and its calories counted for each meal.

At the other extreme are the proponents of another popular hobby—eating too much. There are even more volumes devoted to this pastime. Cookery books can also be aimed at dieters but the best-sellers appeal to the *bon viveurs*, and those who are entertaining friends. They present food in terms of tasty dishes, menus for dinner parties and even exotic banquets. Cookery books may make concessions to the latest health fad but generally their authors seem unperturbed by recipes that require lots of butter, olive oil or cooking fat. They cater for the vast majority of people who still enjoy classical dishes, produced with no thought as to their salt, sugar, fat or calories. Many people have spent most of their lives eating the 'wrong' foods,[12] and clearly it did most of them little harm, although a better diet might have kept a few more of their friends alive for longer and themselves healthier.

The best dietary advice is that which keeps your weight at the right level. Body fat is very noticeable. If you have too much body fat then lose it. But how is this to be done? A bit of chemistry will show you how.

When we diet and take in fewer calories, we make our body draw upon its own store of fat. The best target to aim for is reducing our intake of fat, because triglycerides have more calories than any other type of food. Both fats and oils are equal in terms of calories, and it makes no difference whether it is butter, cream, margarine, olive oil, dripping, polyunsatured spreads or lard. You must cut down on triglycerides if you want to lose weight and stay slim.

The difficulty with diets is sticking to them: a crash diet lasting two weeks tests the limits of most people's endurance. The chemistry of dieting is straightforward, and understanding it may help you to do it successfully. Once you understand the chemistry of food in terms of

[12] Though not all their life, especially if they lived through the 1940s. Britons in the Second World War from 1939–1945, and for a while after, were forced to eat a high fibre diet when all bread was a standard National Loaf. This was deliberately milled to remove less bran, and was unrationed until the world famine of 1947. After that white bread returned.

the energy it provides, you have no-one to blame but yourself if you stay fat. By fat I mean obese, a condition which afflicts growing numbers of people in the West, with over 10% of the population now coming into this category. Obesity is defined by dietitians as a body mass index of 30 or above (see the Box below).

The average man needs about 2500 Calories[13] per day, and the average woman needs 1900 Calories. Exactly how much depends on circumstances. A retired person doing very little may require 200 Calories a day less, whereas people doing strenuous work or sport may need an extra 200 Calories or more. To make the sums easy I will assume the average amount for a man and a woman is 2200 Calories. I will keep the other figures as round numbers as well.

Body mass index

The weight of a person varies with their height and build, and these depend to a certain extent on their sex. To take this into account a system of calculation has been devised called the Body Mass Index (BMI) which can be calculated for each person regardless of whether they are male or female. The BMI of a person is defined as their weight in kilograms divided by the square of their height in metres, so that the average person with a weight of 70 kg and height of 1.80 m will have a BMI of $70/(1.80)^2$ which comes to 22 to the nearest whole number. In the more familiar units of pounds, feet and inches a slightly different formula is required. Divide the weight in pounds by the square of the height in inches, and then multiply this by 700. For example, if a person's weight is 155 pounds and their height is 5 feet 9 inches (i.e. 69 inches in total) then their BMI is $[155/(69)^2] \times 700$ which gives 23.

If your BMI is within the range 20 and 25 you have nothing to worry about as far as your weight is concerned. If your BMI is between 25 and 30 you are too fat and should consider losing some weight, and if it is over 30 you are officially classed as obese. You ought to go on a weight-reducing diet. If your BMI is above 40 you are very obese.

If you reduce your intake of energy to 1000 Calories a day you draw the other 1200 that you need from your own body's reserves.

---

[13] A Calorie is 1000 calories. A chemist would refer to a Calorie as a kilocalorie.

Since your body is not aware you are dieting it supplies the missing calories from its glycogen carbohydrate store in the liver, which it can draw upon straight away. When this reserve is depleted, the body draws upon its fat reserves. Many people fondly imagine that dieting is a direct attack on their flabby bits and are puzzled when these do not immediately start to shrink. Many people also expect their body to shed the fatty tissue which they think appears least attractive, which is around the stomach, hips and thighs. Would that it would.

What does a cut of 1200 Calories mean? You would have to burn up 300 grams (11 ounces) of carbohydrate to provide 1200 Calories. Alternatively your body could consume 140 grams (5 ounces) of its fat. Two weeks on 1000 Calories a day must mean a loss of at least 2 kilograms (4 pounds) if you lost only fat. In fact our body loses carbohydrate, fat and protein, at the rate of about 250 grams (8 ounces) a day. Loss of water from the disappearing tissue will make our weight loss much more than this. A rough guide is that a pound of water is lost for each pound of fat, so that a strict 1000 Calorie diet for two weeks could see your weight drop by an average of a pound a day if you are a man, and slightly less if you are a woman.

There is nothing magical about dieting. You must release energy within your body to keep you warm and active, and that energy must come from fuel provided either by your food or by your own reserves. Obey the rule of a 1000 Calorie diet and you must lose weight—you cannot defeat the laws of chemistry. Hospital dietitians and doctors know that their obese patients are being less than honest when they don't lose weight, while claiming that they are sticking to a 1000 Calorie diet.

Once you have lost weight then the difficult part begins if you are not to put the weight back. Return to your old ways of eating and you could be back to your old weight within another two weeks. Again, chemistry can come to your aid to prevent this happening. You can turn to the substitutes which food chemists have devised to replace high calorie foods. Change to diet drinks, diet soups and diet desserts; buy low-fat snacks, spreads, pickles and dressings; choose calorie-controlled complete meals; and use artificial sweeteners. If you do then you can easily cut 500 Calories a day from your energy intake without any effort. You can also afford to reward yourself with occasional treats like hot buttered toast, a slice of cake, fried egg and bacon, or a pizza. You won't put on weight if you indulge yourself once or twice a week. You cannot put on weight by eating a jam doughnut or a slice of cheesecake occasionally. You only do so if you eat such a treat every day.

Now that you understand the chemistry and energy count, all you need is the will-power. One tip though—don't weigh yourself every day or you will become disheartened when some days you see your weight has gone up despite the diet. The water in our bodies is not something we can control, yet it weighs a lot because we are 60% water. The human body goes through a daily cycle of water retention and loss which reaches its minimum at about midday. The only way to cope with this fluctuation is to weigh yourself once a week on the same day, and at the same time of day.

The other problem with diets is that most humans are rather lazy, and most of us do not have the time to fiddle around weighing bits of food and counting calories. Those who propose a diet plan generally assume that they must make it as varied as possible so that people stick to it. I believe this is the wrong policy to adopt. If the will to lose weight is not there, then no amount of low calorie titbits will keep people on a diet. Those who have to deal with people who are addicted to something realise that they can do nothing until the person really *wants* to be cured. I believe the same goes for dieting. Until you really want to lose weight then diets are a waste of time. But once you have crossed that barrier the rest is easy: you don't need an interesting weight-reducing diet, you need a *boring* diet. Anything less could well tempt you back to eating for pleasure. The other advantage of a boring diet is that you have nothing much to remember, so you know immediately when you have broken it. There can be no trade-off in this diet. Either stick to it and chemistry will *guarantee* you lose weight—or forget it.

Think of the benefits of a boring diet. You save yourself the mental agony of what to choose, because it is the same set of meals every day for two weeks. As an example I give below a diet you you will find easy to remember because the main foods also begin with the letter B. B for boring. For the first three days you may lose weight rapidly as you use up your body's glycogen store (and that will disappear at the rate of about a pound a day), then things will slow down as you start to use your body fat (and that will disappear at about half a pound a day). If you can keep the diet going for an extra two weeks you will lose several more pounds of fat, and then people will really begin to notice. If eventually you find yourself returning to your old ways, then a three-day crash diet will deplete the carbohydrate store before your body has time to start making fat. Again use the boring diet. Once learnt it is easy to remember, but then you have no excuse for not knowing how to lose weight.

The boring diet

A bowl of All–Bran for breakfast, a corned beef or salt beef sandwich[14] and an orange for lunch, baked beans on toast for dinner, and a pint of half-fat milk taken throughout the day. Spread your sandwich with low–fat margarine. Tea, coffee or diet drinks with artificial sweeteners can be drunk any time you want. You can also use an artificial sweetener on your All-Bran. The beauty of this diet is that you can forget the details just remember the Bs: bran, beef, bread and beans. The chemical details are given in the table below (again I am using round numbers to make it easy). A side plate of salad greens and tomato, but with no dressing except salt and vinegar, is also

The boring diet that's guaranteed to work

Meal		Amount	Calories
Breakfast	All-Bran	50 grams (2 ounces)	120
Midday	Corned beef, etc.	30 grams (1 slice)	70
	Bread	2 slices (medium)	200
	Low fat spread	10 grams ($\frac{1}{3}$ ounce)	40
	Orange*	1 medium	40
Evening	Baked beans in sauce	225 grams (8 ounces)	130
	Toast	2 slices (medium)	200
Throughout the day			
	Half-fat milk	400 mls ($\frac{2}{3}$ pint)	200
		Total:	1000

*An apple plus a 100 mg vitamin C tablet will do just as well.

[14] Other protein fillings will do equally well, and you can substitute chicken breast, boiled ham, boiled egg, or canned tuna provided it is in brine.

allowed. The diet is well balanced nutritionally in terms of protein, vitamins and minerals. All it lacks is calories.

The human body is not the fragile flower some would have us believe. People have survived for days trapped without water beneath ruined buildings, have drifted for weeks without food on boats at sea, lived for months in besieged cities with only a few ounces of bread a day, survived a year of starvation in drought-plagued Africa, and even suffered a decade of malnutrition in con-centration camps. Having survived and recovered from their ordeal, they have gone on to live their allotted span of three score years and ten.

Of course food is vital to life, although it can not keep us forever young. There are times in our lives when what we eat is very important—when we are growing up, when we are pregnant, when we are ill, and when we are growing old. For some people, such as diabetics, their diet must be a life-long concern and there are trained dietitians to help. The rest of us can *enjoy* our food, although the sign of a bad diet will be if we end up grossly overweight. Because the pleasure of eating usually wins over our plans to be perfectly fit, we need to watch ourselves, but not obsessively, and we should take advantage of the diet aids that the food chemists have devised.

Several consumer organisations have taken up the theme that we can enjoy a healthier and longer life through better eating. Some people go further and allege that everything from allergies to cancer is due to eating the wrong type of food. Curiously these messages seem to have most impact in those countries where, for the first time in history, most of the inhabitants live out their natural lives and die of the diseases of old age.[15] Can everyone now live longer than the Biblical three-score-years-and-ten, maybe even to four-score years and ten? The best advice is to avoid being overweight. This is much more important than what we eat.

The rapidly growing legions of the elderly consist of people who spent their formative years in eating foods that are now considered unhealthy. When they were young, in the first half of this century, the prevailing philosophy for good eating was to consume lots of cream,

[15] This of course has been brought about by the advances of medical science and public hygiene in eradicating infectious diseases.

milk, and meat, and to avoid so-called starchy foods like bread, pasta and potatoes. In those days poverty and malnutrition, caused by wars and economic depression, were only too real. The poor ate a diet that was largely cheap carbohydrate ('starch'), whereas the rich were better fed and healthier. They ate more dairy products and meat. The inferences to be drawn by comparing the health and longevity of two groups were obvious, but the reasoning, we now realise, was faulty.

You might wonder if one day current food fads will fade in their turn, and the message to cut down on foods rich in cholesterol and animal fats and to eat more fibre will no longer be heeded. This advice is well meaning, and has the support of doctors—but so did the earlier advice. In fact we *should* eat more fibre and less fat, as this chapter has shown. In the final analysis it may matter little what sort of fibre we take more of, or what kind of fat we cut out, and trying to avoid foods with a lot of cholesterol may turn out to be a waste of time.

The chemistry of food has come a long way in the second half of this century, and so has the food industry. Chemists in the industry have been at the forefront of research in this area. Both food chemists and the food industry are regularly criticised by those who campaign for a healthier diet, and I think these criticisms have been a little too harsh. The food chemist, the nutritionist, and the qualified dietitian are all marching in step towards a healthier world. Food chemists are professionals whose first loyalty, one assumes, is to their science. Of course they are not infallible, and the food firms that employ many of them are not run as charitable organisations but for profit, and we must always bear that in mind. In the final analysis though, we are not what we eat, but what we weigh.

CHAPTER FIVE

Painkillers and painful decisions

Thalidomide; painkillers; arthritis and non-steroid anti-inflammatory drugs; opiates; toxicity; risk

THERE are in Nature plants which produce chemicals that relieve pain. However, they did not evolve to provide relief from human suffering, and even though we can extract painkillers from some of them, it is perhaps not surprising to find that these chemicals have side-effects. Indeed, were they to be tested using today's standards for safety in medicines they would fail, and be banned for use on humans. Yet some of these ancient painkillers are still the most effective we have. They are given to relieve the intense pain of accident victims, or of those who have undergone surgery. They are used to make life bearable for people in the last stages of painful diseases such as cancer. The drugs in question are the opiates **morphine**, **diamorphine** (better known as heroin), **codeine** and **hydromorphone**.[1]

Merely to deaden pain is simply to treat the symptoms without dealing with the underlying cause of that pain. For this reason we are perhaps less willing to take risks with painkillers. We accept that healing treatments involve some degree of risk in the form of adverse side-effects, and these we are prepared to live with in the hope of securing long-term benefits to our health. On the other hand, ending the pain of a headache or an inflamed muscle is purely a short-term

---

[1] A reminder that when the name of a substance appears in **bold** type, it means that there is more about its chemistry in the Appendix.

gain, lasting a few hours at most, and if we take something to ease that pain we want it to be free of risk.

If a person demands *absolutely* no risk at all, then they are demanding too much, and would be best advised to seek relief in treatments that do not involve taking a pharmaceutical compound. They should opt for something like homeopathic medicine or aromatherapy. If these work, then they probably do so by relieving psychological stress, which might well be the origin of the pain. Most people who want relief from pain turn to over-the-counter[2] drugs such as **aspirin** or **paracetamol** when they have an ache in their head, back, joints or muscles. Many old people also take these general purpose painkillers as prescribed by their doctor, and they take them every day to deaden the pain of arthritis. All these painkillers carry a risk, and naturally this has led to attacks on them, not least by those who offer alternatives or who see them as 'chemical' cures as opposed to 'natural' cures.

Should a warning signal flash through our mind every time we swallow a headache pill? Perhaps it should, because there are few times in life when you are taking into your body such a large amount of a potent chemical. The chemists of the pharmaceuticals industry may have transformed the quality of life for billions of people, but they have sometimes made terrible mistakes. The thalidomide tragedy of deformed babies is one of which we are often reminded. I shall begin this chapter with the thalidomide story and tell how this drug is still being used. I will then go on to explain the chemists' fight to control pain, and finally end the chapter with a discussion of the vexed question of how much risk we should be prepared to accept when taking painkilling drugs.

Thalidomide

A woman who is expecting a baby can enjoy her condition for much of the nine months of pregnancy, although many experience morning sickness during the early months. While this may be perfectly natural it is inconvenient, and can sometimes be very unpleasant. At one time

---

[2] Pharmaceuticals that are available for self-medication are generally known as over-the-counter drugs.

there was little a doctor could do but give reassurance, although this was not much help if the woman was also trying to lead a normal life and do a full-time job.

Then suddenly there was thalidomide. In 1958 this drug was launched in Germany in a blaze of publicity by the company Chemie Grünenthal, who even felt it was safe enough to be sold without prescription under the brand name Contergan. Within a few years it was available in over 40 countries. The liquid form of thalidomide quickly gained a reputation in Germany as the babysitter's friend because it was superb at calming young children and sending them to sleep. Chemie Grünenthal were so confident that Contergan was safe that it was advertised with a picture of a child taking a bottle from a shelf with the claim: 'completely harmless, even for infants'.

There were some side-effects—a few patients complained that thalidomide made them constipated and slightly dizzy—but these kinds of reports are to be expected for any drug, and mean very little. In the UK, thalidomide was called Distaval and was prescribed as an anti-depressant, and it was also sometimes given to control morning sickness. It was much safer than the other anti-depressants that were available, such as the barbiturates, and accidental overdoses were not lethal—indeed, an adult could take as much as 350 grams (12 ounces) of thalidomide without its being fatal. It seemed the perfect sedative, and it stopped the morning sickness.

Doctors and patients were unaware of a dreadful side-effect of this drug—it was a teratogenic compound. This is the technical term for a substance which causes a developing foetus to become deformed. A woman who was prescribed thalidomide in her first weeks of pregnancy was at risk of giving birth to a limbless child. A typical thalidomide baby could have an arm missing, and just a tiny hand protruding from its shoulder.

The first inkling that something was wrong came in 1961 when a young doctor, William McBride, who worked at the Crown Street Women's Hospital in Sydney, Australia, first noticed an unusually high number of babies being born with tiny arms and legs, a deformity that was not unknown but was normally extremely rare. When he made enquiries of the mothers, he discovered that they had all taken thalidomide during pregnancy. McBride wrote to the medical journal the *Lancet*, pointing out this coincidence. His suspicions proved to be well founded, and when they were followed up the link with thalidomide became evident. As a result the drug was withdrawn in most countries.

The publicity and sympathy for the victims of thalidomide, and the large financial settlement which they were given, highlight the risks that are attendant on all new pharmaceuticals. Yet Chemie Grünenthal had a superb product in thalidomide, and what they did not realise was that while half the molecules in thalidomide were safe, the other half were anything but. The two forms of thalidomide are identical except for one minor detail: they come as left- and right-handed pairs.

Gloves, cars, screws and molecules have one thing in common: they can all be made in two versions—left-handed and right-handed. We never think twice about gloves coming in pairs, and we know that cars built for Australia, Britain and Japan have the driver's seat on the right side, while for most of the rest of the world the driver's seat is on the left side. Screws are less obviously of two kinds, and most people would not know, just by looking at it, whether a screw had a left- or right-handed thread. Molecules are like screws in that we would not instinctively know by looking at one that it had within its structure the ability to be right- or left-handed. Not all molecules are like this; it is a feature which generally comes with size and complexity.

Chemists talk of such pairs of molecules in terms of one being the mirror image of the other. They label them not as left or right, but as S (from the Latin *sinister*) or R (*rectus*).[3] R-thalidomide was fine, but the S-thalidomide could interfere with the development of the foetus and deform it. In the form in which it was manufactured, thalidomide consisted of both R and S molecules in equal numbers, and back in the 1950s it would have been impossible to separate them on a commercial scale, even if they had known that this was necessary.

Today chemists can do this separation relatively easily, and now they *must* do it, because all new pharmaceuticals have to be tested separately as both their R and S versions. If this had been a stipulation back in the 1950s then doctors would still be prescribing R-thalido-

---

[3] An alternative system which is still widely used has the letters D (from the Latin *dexter* for right) and L (*laevo*, left).

mide as the remarkable multi-purpose drug that it is. In those days, however, chemists had neither the skill to separate left- and right-handed molecules easily, nor did they have chemical reactions that could make the one without the other. Advances in chemical research over the past 40 years have brought such achievements about, but this has still not been done for thalidomide, presumably because the cost would never justify the commercial risk involved in trying to reintroduce a drug with such a negative public image.

Yet thalidomide is still being manufactured in Brazil, using the original recipe, and is given to treat leprosy. Provided it is not taken by a pregnant woman there is no risk. Thalidomide is remarkable at dealing with the adverse effects which can follow any treatment that involves destroying bacteria in large numbers, which is what happens in leprosy. It is possible to prescribe thalidomide in some countries such as the UK, and this is done to treat patients where such conditions apply. Thalidomide is still being investigated because it has the potential to deal with inflammatory conditions by suppressing the immune system. This has alerted researchers to the possibility that it might also alleviate AIDS, rheumatoid arthritis, and even prevent the rejection of organ transplants. Were this to be proved it might one day justify the manufacture of the *R*-form.

Painkillers

There are natural painkillers within our brain called enkephalins. These consist of five amino acid units joined together as a short peptide chain, and they are powerful painkillers produced in response to traumas and with effects similar to morphine. They are derived from longer peptides called endorphins, which are made up of 80 or so amino acid units.[4] Endorphins come to the aid of the brain when the body has to cope with intense pain experienced over a long period, but they can not be produced to relieve such pain immediately, and this is why we seek quicker-acting painkillers.

There are basically two kinds of painkilling drugs: those which operate at the site of the pain and stop the damaged tissue sending

---

[4] The name endorphin is derived from the phrase 'endogenous morphine', meaning morphine produced from within.

signals to the brain, and those which act in the brain and prevent the incoming signals from registering as pain. The latter, which act on the central nervous system, are more likely to have side-effects such as affecting a person's mood, and some are addictive. The former, which act peripherally, may have side-effects as well, but sometimes they may even be beneficial in actually treating the cause of the pain itself.

Before chemistry and the pharmaceutical industry developed, the only source of painkilling chemicals was plants. Hippocrates, the ancient Greek pioneer of medical science, had recourse to some such remedies; for instance, he recommended an infusion of willow bark to ease the pain of childbirth. If this bark was that of the white willow then we know that this advice would have been of some use, because this tree produces salicin which on boiling in water is converted to **salicylic acid** which is a painkiller. This chemical is called after the willow, whose scientific name is *Salix*. Although salicylic acid deadens some pain it can also cause pain, because it irritates the stomach lining to the extent of causing bleeding.

Today there are four commonly used, over-the-counter painkillers: aspirin, paracetamol, codeine and **ibuprofen**. They are taken in their millions every day, and some countries favour one type over another. For example Britons prefer paracetamol-based tablets, while the second most popular is aspirin, the painkiller that helped launch the modern drugs industry a century ago. Aspirin is still the number one in the USA. Codeine has been around a long time and is also still popular, but is now only sold combined with other painkillers for reasons we shall see. The newest over-the-counter painkiller is ibuprofen which is slowly gaining in popularity.

Aspirin

It was to be another 2000 years before Hippocrates' advice about willow bark was taken further. In the eighteenth century the Reverend Edmund Stone, an English parson living in the Cotswolds, was trying to find a local plant whose bark might have the same effect as cinchona bark that was imported from Peru. This Peruvian bark contains **quinine**, which is a natural chemical that was known to ward off fevers, and especially malaria. Parson Stone experimented with the bark from the white willow, and gave an infusion of this to 50 people who had fever and found they were much improved. He reported his findings to the Royal Society of London in 1763.

In the 1820s an Italian chemist, Raffaele Piria, isolated the chemical salicin from willow bark, and from it he made salicylic acid. It was later discovered that this acid could also be extracted from the wild flower meadow-sweet (*Spiraea ulmaria*).[5] It is from this plant that aspirin gets its name (a+spiraea), and the word was coined by the firm Bayer for the derivative, acetylsalicylic acid. This had first been made around 1850 by a Frenchman, Charles Frederic Gerhardt, but although it was known to be a painkiller it burnt the mouth and stomach of those who tried it, and it tasted unpleasant. It later transpired that these side-effects were caused by impurities in the product.

In 1893 Felix Hoffmann and Henrich Dreser, chemists working for Bayer, found a way of purifying acetylsalicylic acid and obtained it as a white powder. It was worth taking the trouble because by doing so they produced one of the most successful proprietary drugs ever made, and one considered safe enough to be sold direct to the public. Americans swallow 50 million aspirin tablets a day, which is almost two billion a year. Aspirin is not risk-free and would not be passed for use under modern testing methods. It causes stomach inflammation in some people, and it slowly reverts back to salicylic acid which is even worse. For this reason old aspirin tablets should never be taken.

Aspirin is regarded as more than just a painkiller. At various times in the 100 years since it was first marketed it has been the standard treatment for virus infections (such as fevers, colds and 'flu) and for arthritis (once commonly referred to as rheumatism). It is a cure for none of these, it just makes them more bearable by reducing the aches and pains they cause. Aspirin is still bought and taken for these complaints under a variety of trade names, such as Anacin, Anadin, Aspro, Disprin, Ecotrin, Excedrin, etc., and of course as its best known version, Alka Seltzer. Nor is this plethora of names limited only to aspirin: it has become an integral part of the drug industry (see the Box).

---

[5] Salicylic acid itself is a simple organic molecule made industrially from phenol and carbon dioxide.

What's in a name?

It has been said that most human illness is psychological in origin, and so perhaps we should not find it so surprising that psychology plays an important part in the effect of pharmaceuticals. We all know about placebos, the capsules of glucose or some such, which are given in double blind tests, and which we suspect doctors sometimes give to their more troublesome patients who are simply seeking attention. In fact it would be unethical for a doctor to do so. What doctors may opt to do instead is to prescribe the same medication under a different name. Most pharmaceuticals have many names.

Every drug has at least two official names: the chemist's systematic name, which may run to several lines of text, and the agreed *generic* name which is approved by international agreement. The chemical name, with its curious italics, brackets, Greek prefixes, symbols or numbers, is unreadable and easy to get wrong, especially by those who are not familiar with them. The generic name is usually simple, short and memorable. Thus we have *N*-acetyl-*para*-aminophenol as the chemical name, and paracetamol as the generic name.

In addition there are the trade names, of which dozens are in use for this one drug alone. We may be persuaded by heavy advertising always to buy Panadol, Tylenol, or whatever name it is given in the place in which we live; and in just one country, the UK, doctors can give paracetamol as Alvedon, Cafadol, Disprol, Paldesic, Panaleve, Salzone, and Tylex. Some of these are mixtures with other compounds.

Doctors can prescribe a drug under its trade name or generic name, when it will be cheaper. But is the cheaper alternative really costing us dear in the long term? Companies who make generic drugs generally do no research. On the other hand the big pharmaceutical companies need their higher prices to fund the huge costs of research into new drugs. Sadly, they need to spend even more on promoting, protecting and improving their existing drugs, than on research. What the research-based pharmaceutical companies require are longer patent rights, starting from the date the drug is finally cleared for use, rather than from the date of its discovery. Until such times they need the support which a well-known brand name can provide. That's what's in a name.

Tablets of Alka Seltzer also contain **citric acid** and **sodium bicarbonate** as well as aspirin. The bicarb reacts with the aspirin to form its sodium salt, thereby making it soluble in water and quicker acting. It also reacts with the citric acid to generate bubbles of carbon dioxide. The citric acid gives the final drink a pleasant citrus-like flavour.

Soluble forms of aspirin can be made by converting the acetyl-salicylic acid, which is not very soluble in water, into its sodium or calcium salt, which is soluble. When dissolved in water these give a clear liquid which some find reassuring to drink and easier to take than swallowing a tablet. However, once aspirin is in the stomach the acid conditions there immediately turn it back to the insoluble form, although this is now as very fine crystals. These are slightly less irritating to the lining of the stomach than particles of a large tablet. Yet even they may be too irritating for some patients, in which case a doctor can prescribe a clinical form of aspirin, called enteric aspirin, in which the tablets have a coating that resists the acid of the stomach but dissolves once the tablet passes into the small intestine where the conditions are alkaline.

Although aspirin has been used for a long time it is not without its more serious risks. For some young children aspirin can be fatal, and if they are given it to treat the symptoms of a viral infection, such as 'flu or chicken pox, they can develop Reye's syndrome. For this reason children under the age of 12 should never be prescribed it, and in the USA aspirin packs must carry the warning: 'Children and teenagers should not use this medicine for chicken pox or 'flu symptoms before a doctor is consulted about Reye's syndrome, a rare but serious illness'.

Despite its disadvantages, aspirin is still a remarkable drug and much more than just a painkiller. It is said to ward off heart disease and prevent thrombosis, and it has even been mooted as a possible treatment of cancer, cataracts and senile dementia. It is used in the treatment of people who have suffered a heart attack. If you suffer no stomach pain with aspirin then you might be well advised to take a small aspirin tablet (100 mg) each day. Aspirin inhibits the formation of those chemicals which cause blood platelets to aggregate together, which is what starts a blood clot. This may also be why it might help in senile dementia, because it keeps up the flow of blood to the brain. Aspirin is also thought to be able to destroy free radicals, those super-active natural chemicals that are formed within the body and which are thought to initiate cancer. Aspirin can also prevent cataracts because it counters the proteins that make the lens of the eye opaque.

Aspirin is normally sold as 300 mg tablets and can safely be taken at a rate of two or three every four to six hours to a maximum dose of a dozen tablets or about 4 g per day. A single dose of 10 g (30 tablets) is life-threatening for an adult because it makes the blood too acidic. The body tries to cope by rapid breathing to dispel CO_2 and thereby reduce acidity, and by boosting the action of the kidneys, which leads to dehydration. If the acidity cannot be corrected by natural means, tissue damage occurs and eventually death.

How does aspirin work? It interferes with an enzyme that makes **prostaglandins**. We need these chemicals to regulate digestion, kidney function, blood circulation and reproduction. Prostaglandins also generate pain signals. They are released in many injuries and diseases, and it is because of them that we suffer inflammation, pain and fever. This was discovered in 1969 when John Vane and a group of workers at the Royal College of Surgeons in London, England, proved that aspirin prevented the synthesis of prostaglandins in damaged tissue. Vane shared the 1982 Nobel Prize in Physiology or Medicine with Sune Bergström and Bengt Samuelsson of the Karolinska Institute, who had also studied the role of prostaglandins.

There is a whole sequence of chemical reactions which lead to the appearance of prostaglandins, and the first is the liberation of the polyunsaturated fatty acid called **arachidonic acid** from a cell membrane. This acid reacts with oxygen to form chemicals such as cyclic prostaglandin endoperoxides that trigger inflammation. Aspirin blocks the enzyme which controls this reaction and so prevents the chain of events which starts with a broken cell membrane and leads to painful prostaglandins being produced.

Arthritis and the non-steroid anti-inflammatory drugs

Some drugs relieve inflammation, such as steroids like cortisone, but are not painkillers, and some relieve pain, such as morphine, but are not anti-inflammatory, and these latter types we will deal with later in the chapter. There are some drugs which do both, and these are the non-steroid anti-inflammatory drugs (NSAIDs). Aspirin is one such drug, and it used to be the first line of defence in the treatment of arthritis.

Rheumatoid arthritis occurs when the body mounts an immuno-logical attack on itself, in this case on the connective tissue between the joints. It can deform the body and is a crippling illness. There can be periods of remission and most patients improve with care and exercise. Osteoarthritis is less damaging in that it does not deform the body and is generally considered to be part of growing old. It affects the joints of the hips and knees, and is made worse if the victim is overweight. According to the Arthritis Foundation in Atlanta, about 75 people per 10 000 in the USA are afflicted with rheumatoid arthritis, and it is twice as common in women as in men. Osteoarthritis affects nearly 600 per 10 000, which is over 16 million Americans. Again, two thirds are women.

Osteoarthritis is caused by wear and tear of the cartilage which surrounds the bone at the joints. Cartilage has no blood supply to bring in the chemicals it needs, but it is 70% water and it uses this to do the job instead. As joints move this water is squeezed out of the cartilage, carrying unwanted wastes with it, and then within fractions of a second the joint relaxes the pressure and water seeps quickly back bringing in fresh nutrients.

Clearly, the more exercise we do the better our joints will be, but as old age creeps on not only do we exercise less, but our cartilage also begins to age and, like a rubber seal between two metal joints, it weakens and begins to leak. As it breaks down it begins to produce arachidonic acid which then leads to the production of prostaglandins which cause inflammation. This in turn makes a person less likely to move the joint, so the condition intensifies in a vicious circle.

As the cartilage fails, the bone beneath tries to compensate by growing harder, as the joint now has bone grinding against bone. The harder bone, however, is more likely to crack if we fall or bang it hard. These cracks then grow more hard bone to compensate and the joints start to deform. Scientists may one day find a chemical that will lubricate the joints, or even stimulate the body to repair its own worn-out cartilage. Until that day we have to rely on drugs that limit the damage and the pain.

There was a period after the Second World War when steroids were the recommended treatment for arthritis. These check the disease but can have serious side-effects, and this led to the search for non-steroid drugs to do the job. Several have been discovered, but these were not a cure for arthritis—they merely suppressed the symptoms and made the condition easier to live with. Recent pharmaceutical advances have gone some way toward a cure, and doctors can now prescribe drugs such as penicillamine, hydroxychloroquine (which is also used to treat malaria), sulfasalazine, and even ones containing gold. If you suffer from severe arthritis and you are given **auranofin** (the trade name is Ridaura) then you are taking a molecule which contains an atom of gold at its heart. This drug is effective in suppressing rheumatoid arthritis. It can be taken by mouth, rather than being injected as this type of drug generally is, and it has fewer side-effects.

Meanwhile most people who suffer from arthritis are helped with the so-called non-steroid anti-inflammatory drugs (NSAIDs). The sales of these in the USA amount to around $2 billion per year. In Western countries there is a potential market of around 50 million older people who can easily afford treatment, which may explain why drug companies are prepared to invest so much in the search for a cure for arthritis.

There are several NSAIDs now available to doctors, so that if a patient cannot tolerate one of them there are alternatives that can be prescribed. Doctors will usually start with ibuprofen or paracetamol (see below), and if these provoke an adverse reaction then they have a repertoire of about 25 others to try until they find one that can be tolerated. It really does help for there to be alternative NSAIDs, and doctors prescribe the newer NSAIDs because these are slightly better than the older varieties, which are in any case out of patent and earning no royalties for the companies that discovered them. In fact 25 seems somewhat excessive, but there are still more awaiting approval. NSAIDs differ generally in their acidity and their ability to penetrate to various parts of the body. What suits one person will form ulcers in another, but why this is so remains a mystery. Two of them are even sold as over-the-counter drugs and are also prescribed for arthritis sufferers, and millions of people take them as general painkillers.

Paracetamol

Paracetamol, like aspirin, is a chemical derivative of phenol. It too is sold under many names, such as Tylenol in the USA and Panadol in the UK. It has two advantages over aspirin: it is absorbed through the lining of the stomach and so acts within about ten minutes, whereas most aspirin is absorbed through the gut and may take an hour or more to start working. Nor does paracetamol inflame the stomach lining. The pain-killing properties of paracetamol came to light when **acetanilide**, a derivative of benzene, was mistakenly added to a patient's medicine and proved to be an excellent pain-killer. Acetanilide is rather toxic, and the body detoxifies it by converting it to paracetamol. Clearly the benefits of acetanilide could better be obtained by prescribing paracetamol directly. Like aspirin, paracetamol works by blocking the enzyme needed for making prostaglandins.

Paracetamol is also a drug we can misuse. Two paracetamol tablets will give temporary relief from muscular pain or a headache, but 22 tablets will make that relief permanent—in Heaven. Yet two 500 mg tablets can safely be taken four times a day by an adult, week after week, month after month, year after year, provided the patient is of average build and weight and in reasonable health. Small and under-weight people have been known to die of paracetamol poisoning having taken only ten tablets in a 24-hour period.[6]

Others at risk of paracetamol poisoning are would-be suicides, alcoholics with already damaged livers, and people suffering the condition known as bulimia nervosa. Bulimia nervosa generally afflicts young women who feel a compulsive urge alternately to starve themselves and then to over-eat. Some take paracetamol to make themselves vomit.

Those who attempt to commit suicide with paracetamol often succeed, even though they change their mind, or are found in time. They can be pumped out, put on an artificial support system and given blood transfusions, but this often cannot save them. Paracetamol has damaged their liver beyond repair and the

---

[6] Some people may inadvertently take dangerously high doses of paracetamol if they have a cold and take a combination of tablets, cough linctus and 'night time' medicines, all of which might contain paracetamol.

temporary recovery will only last a few days. They will die after about a week. Around the world paracetamol kills over two thousand people each year.

Although paracetamol is good at blocking pain receptors, our body treats it as a toxin to be disposed of as quickly as possible, and this is why its toxicity appears so puzzling. Our body has two ways of removing paracetemol—the most common route is to use enzymes to add a sulfate group to the molecule. This makes it very soluble in water so that it can be filtered out by our kidneys. Alternatively our body's enzymes can use glucuronic acid, a derivative of glucose, to add a carbohydrate to the paracetamol, and this too will help its removal. While the amount of paracetamol in our body is low, as it is with the recommended dosage, then these enzymes can cope quite easily.

When we take too much paracetamol the body tries its best to remove it with sulfate or glucuronic acid, but when these cannot cope it relies on oxidation. This is risky because it converts it to a deadly poison which then has to be quickly removed by the liver, which uses **glutathione** to do the job. However, glutathione is not easily replaced by the liver and is soon used up. The poisonous compound then attacks and destroys the liver itself. Ten hours of assault by the poison and the liver is beyond repair.

All this can be prevented with an injection of a compound called **N-acetylcysteine** which can act just as well as the liver's own glutathione. Accidental overdosing on paracetamol is not dangerous provided you get an injection quickly. It could even be made impossible by adding methionine to tablets of the painkiller. This simple sulfur-containing amino acid increases the amount of glutathione in the liver naturally. There are now plans to market a safe form of paracetamol with methionine.

Ibuprofen

This NSAID was discovered by a group of chemists at the drug company Boots, in Nottingham, England, who were trying to improve upon aspirin. The idea was to retain its painkilling and anti-inflammatory action while reducing the irritation to the stomach. At first they found that changes to the molecule would indeed make it more potent as a drug, but its toxicity increased as well. Sometimes they found compounds that had ten times the painkilling ability of aspirin, but there was always some drawback. For example, one variant caused people to come out in a rash.

After making hundreds of versions they eventually hit upon ibuprofen. It was over 20 times better than aspirin as a painkiller, and it controlled inflammation and reduced fever. The first clinical trial took place in 1966, and doses of 200, 400 or 600 mg showed no adverse side effects. Ibuprofen was so successful that it became an over-the-counter drug in the UK in 1983—Boots sold it under the trade name Nurofen. It also went on sale in the USA as Nuprin, Motrin and Advil. A daily dose of three 400 mg tablets is the accepted rate of prescription, but twice this amount can be given under certain circumstances.

Of course ibuprofen did not agree with everyone, and perhaps it was inevitable that an alarm bell was to be sounded. Scientists at Johns Hopkins University Medical School in Baltimore published a paper in *Annals of Internal Medicine* in 1990 which claimed that 12 women on ibuprofen suffered kidney failure when taking the drug. In fact they were all suffering chronic kidney disease before treatment with ibuprofen started. Was ibuprofen to blame for worsening their condition? According to Boots such patients probably should not have been put on the drug anyway, and the same is true of people with peptic ulcers. The number of women involved was actually quite small compared to the millions who had obtained relief with ibuprofen and who had come to no harm. The media sensed a scare story but quickly lost interest once the facts were explained.

Not all drugs escape the media's attentions so lightly. Another NSAID, Opren, was superb at relieving the symptoms of arthritis sufferers, and they watched in dismay as a media campaign in the UK led to its withdrawal. Again, a few unexplained deaths of some very frail people who were given Opren set off the alarm. We will return to the Opren affair later, because it illustrates the choices we as a society sometimes have to make.

Opiates

The opiates[7] are the most powerful of all painkillers. The opium poppy, *Papaver somniferum*, is the source of these, and codeine, morphine, diamorphine and hydromorphone are the best known derivatives. All are effective painkillers, with side effects that include constipation, and two also cause addiction. Diamorphine and morphine carry this risk, and may take as little as a week to become established.[8] Hydromorphone may also come into this category.

However, it is the painkilling features of the opiates on which we want to concentrate. One of the great triumphs of modern chemistry has been the discovery of new molecules that relieve pain, but few discovered so far can compare with the opiates. The raw material can be collected by making cuts in the unripe poppy seed capsules, from which a milky juice exudes which dries and can be scraped off. This is the traditional form of opium. For the pharmaceutical industry the opium is harvested mechanically. Its main component is morphine, a complex organic molecule which acts on the central nervous system by blocking the so-called opioid receptors which register the sensation of pain. If you take some opiates, such as morphine or diamorphine (heroin), even though you are not in pain, then you experience their well-known narcotic effects.

Morphine works because it is chemically like the body's own painkillers. However, these take many hours to build up in the brain, whereas morphine can be injected and works within seconds, ending even the most unbearable pain. Sadly, it is addictive. Modifying the molecule slightly, by replacing one of its hydrogen atoms with a methyl group, got over this problem. The new compound was called codeine, which has lost its power to produce addictions but it is much less effective. However it is safe enough to be included in many over-the-counter painkillers.

---

[7] The word opium comes from the Greek *opion* meaning poppy juice.

[8] **Methadone** can be used to help people break with heroin; this takes about 30 days of regular use to produce addiction. Methadone is most active as the left-handed molecule (see page 121).

Morphine can be changed into an even quicker-acting painkiller by replacing two of its hydrogen atoms with acetyl groups. This is diamorphine, derived from the name diacetylmorphine, and it can now penetrate the brain more easily than morphine. When it gets there it is quickly reconverted into morphine by losing its acetyl groups. Diamorphine has the power to relieve terminal cancer patients of their pain, and is prescribed in Europe for these people in their last few months of life. To them it matters little that it is even more addictive than morphine, and its narcotic effect may even make them feel better—for a while. In the USA the opiate hydromorphone has been approved for use, and can be prescribed to relieve the suffering of cancer patients. It is the most potent opiate painkiller of all, as the table below shows.

The potency of the opiate painkillers

Opiate	Potency*	Solubility**
Codeine	60 mg	20%
Morphine	20 mg	5%
Diamorphine	10 mg	50%
Hydromorphone	2 mg	25%

*This is the maximum safe single dose. The smaller the dose the more potent is the molecule.
**This is the solubility in water of the phosphate salt of codeine and the hydrochloride salts of the others. The more soluble the drug, the easier it is to administer, and therefore the more effective it is.

In 1992 a new painkilling molecule was reported that was even stronger than morphine. It is produced by an Ecuador tree frog *Epipedobates tricolor*. John Daly at the National Institutes of Health at Bethesda, Maryland, extracted it from the frogs in tiny amounts and called it **epibatidine**. It turned out to be 200 times as effective as morphine in standard pain tests on rats. Not only that, but the molecule is also relatively simple and is an organochlorine compound. Whether epibatidine will open a new era of non-toxic, low-dose, non-addictive, over-the-counter painkillers remains to be seen. Before an epibatidine-type molecule is used it will have had to undergo a battery of tests to ensure it is effective, and that the risk to life of those who take it is as low as possible. As we shall see, the chances are that it will fail such a test.

Toxicity

We know that it may only take one molecule to attack the DNA in a cell, and to cause that DNA to mutate. This mutant cell can eventually multiply into millions until it kills us of cancer. Although this is true, it is not the whole truth. Every one of the billions of cells in our body has its DNA damaged by radioactivity or by chemicals, by which I mean natural background radiation and by natural chemicals. Evolution produced the living cell and created a wonderful thing, endowing it with ways of checking itself, so that when it finds itself with broken or mutant DNA, it can remove it or repair it. We may of course expose ourselves to more radiation or dangerous chemicals than our defences can cope with, and end up with a damaged cell that escapes detection, or one that our repair system cannot cope with. Then we are in real danger, but that is such an unlikely event for most of us that we can ignore it for most of our life.

Oxygen gas, which we think of as so essential to keeping us alive, is in fact far more dangerous than many 'chemicals', and yet we take in massive amounts of oxygen each day. Oxygen will attack anything given a little help, and it can attack various parts of cells, because it forms dangerous chemicals called free radicals. These are molecules with a loose electron and they can react with anything. But we do not need to worry about oxygen behaving like this, because our bodies have lots of antioxidants on hand to destroy such free radicals. Likewise we also have a superb defence system to protect us against the toxins and unwanted chemicals that we take in with our food quite naturally.

The plants and animals that we eat were not designed to be safe for us to dine on; in a way we had to design ourselves around them. Of course there are some natural poisons so dangerous that they kill us if we eat a tiny amount, and there are lots of substances in our diet which are toxic, but which we encounter only in amounts too small to be lethal. We still need to detoxify them, and our body does this by modifying them slightly to make them more soluble in water (as we saw with paracetamol). In this way our kidneys can filter them out and get rid of them.

The same action is taken against all threatening molecules, and explains why we do not get cancer even when we take in chemicals that can be shown to cause cancer in laboratory rats. We can prove that a chemical causes cancer by exposing a rat to it in huge doses. We can be exposed to the same chemical, but in much lower doses of course, day after day, all our lives and yet it never causes cancer, because day after day our defences cope easily with it. It may also be that unlike the rat's genes, our genes have in-built protection anyway. This is much more likely to be the case than perhaps we realise.

We could ask the same question about the human response to medicines: why should one person be cured by a dose of a drug while another person suffers side-effects, and maybe even not tolerate the drug at all? The answer to this idiosyncrasy lies in our genes, and in how these have equipped us to detoxify chemicals. One way to detoxify is to modify an alien molecule by adding an acetyl group of atoms to it, and so make it more soluble in water. (You can see the effect of acetyl groups by comparing the solubilities of morphine and its acetylated counterpart, diamorphine, in the table on page 134.) The enzyme which enables us to do this acetylating naturally is caused by a single difference in one chromosome.

Asian people are generally fast at acetylating, Americans and Europeans less so. A poor ability to do this may put some people at risk. Earlier this century workers in the dye industry were exposed to benzidine and *ortho*-toluidine, both of which can cause bladder cancer. Those who were good acetylators were less at risk than their poor-acetylating colleagues. Of the 23 men who died of this condition in Britain, 22 were found to be slow acetylators, and so their bodies were exposed to higher levels of these chemicals for longer.

Another important property of a painkilling drug is its duration in the body. How long will its effect last? It is important that it should last a long time. The dose of a drug that you are prescribed can vary from less than 10 mg, about the size of a pinhead, to 1000 mg, the size of an Alka Seltzer tablet. The reason is partly related to how long it takes your body to get rid of what it regards as an alien chemical—even though it is there to help. It may take days to remove the last molecule of a substance from the body even though the drug's effectiveness has long since worn off, so that the time for complete removal from the body is no real help in deciding the dose.

When comparing the retention of a drug in the body, scientists refer to its half-life. This is the time it takes for the body to remove half the given dose. If something has a half-life of ten hours, then after ten hours the amount will be half the original dose, after another ten hours it will be halved again (in other words down to a quarter of the original), in another ten hours it will be halved yet again (now an eighth of the original), and so on. We should still be able to detect a trace of the original drug even after 100 hours, when the amount would be about a thousandth (0.1%) of the original.

Even when we are talking about a system as complicated as an individual's body chemistry, we can still use the concept of half-life, although it will of course vary from person to person. For example, it will depend on how well their liver and kidneys are working, but the list in the table below gives the usual half-lives of some chemicals in the average healthy human body.

The half-life of chemicals in the human body

Drug	Half-life
Aspirin	15 minutes*
Caffeine	5 hours
Cocaine	45 minutes
Valium	18 hours
Morphine	3 hours
Nicotine	2 hours
Streptomycin	5 hours

*Fifteen minutes seems too short, but aspirin works by being converted to another chemical which has a much longer half-life.

About half of all chemicals tested are carcinogenic, but this does not mean that they automatically cause cancer. Every mouthful of food we eat contains some arsenic and arsenic is a known carcinogen, but we happily survive this daily onslaught because our body can cope with arsenic and may actually need a small amount of it. Because a substance causes cancer when fed in large doses to animals, it does not mean that it will cause cancer when given in low doses to humans. We have to swamp the natural defences of a laboratory mouse day after day before we can show whether it can trigger off a cancer in the poor animal.[9] But if you carry out such tests, and a cancer results, then in the USA the chemical is banned by the Food and Drug

---

[9] Such mice are specially bred to be particularly prone to get cancer anyway.

Administration (FDA), who are regarded as the world authority on safety.

The FDA came into being as a result of a drug disaster, and it justified its existence 20 years later when its lengthy procedures happily prevented the thalidomide tragedy from happening in the USA. This success, though, was more by accident than design. The case which brought the FDA into existence was the first antibacterial drug, **sulfanilamide**. In December 1932 a chemist called Gerhard Domagk, working for the German chemical conglomerate IC Farben, patented an azo-dye called Prontosil. Safety tests were carried out on rabbits and cats and they were not harmed by it, but tests on infected mice revealed something rather unusual. Twelve mice which had been injected with streptococci bacteria that would normally kill them within two days recovered when given Prontosil. Tests at a local hospital confirmed its efficacy, and Domagk even used his new compound to cure his daughter of an infection.

What Prontosil did was to produce sulfanilamide in the body, and this is the key chemical that was effective, not the dye itself. Prontosil could not be patented as a therapeutic agent because the sulfanilamides had been recognised as having antibacterial activity in 1919. We now know how sulfanilamide works: it blocks an enzyme that the bacteria need to make folic acid, which is essential for them just as it is for humans. But whereas we take in all the folic acid we need from our food, bacteria have to make all theirs themselves. Giving a dose of sulfanilamide to a human who was infected with bacteria would stop these spreading, and hold them in check until the body's defences could produce its own antibodies to kill the bacteria.

Sulfanilamide was launched in 1936, and for the first time doctors could prescribe a successful treatment for conditions such as bacterial pneumonia in children. In Britain it was known as M&B after the firm May & Baker which produced it, and it was given in tablet form.[10] Tablets are difficult for children to take and a liquid is much easier to swallow, but there was a problem with sulfanilamide. It was not very soluble in the normal solvents used to administer medicines, which in those days were either water or alcohol. It was soluble in another kind

---

[10] I was given M&B when I had pneumonia as a child, so I can testify to its effectiveness.

of alcohol, **ethylene glycol**, a liquid normally used as antifreeze, and that was how it was supplied for use as a so-called 'elixir'.

In 1937 doctors began to report to the American Medical Association that some patients unexpectedly died when they were treated with sulfanilamide. Supplies of the drug were recalled but 107 people died, mostly children, before all the samples were traced. Tests by a Dr Francis Kelsey showed that the cause was not the drug but the solvent. The tragedy of the elixir of sulfanilamide led to the 1938 Food, Drug and Cosmetic Act, and this stated that henceforth the FDA must approve all products before they are put on the market.

Drug companies of course carry out very stringent tests before they hand a new drug over to the FDA for approval. Even so, about a quarter still fall at the first fence, the so-called phase 1 trials which involve up to 100 patients. Another quarter will fail the phase 2 trials, which involve several hundred people. And a third quarter will fail the phase 3 trials which involve thousands of volunteers. Only about one drug in four emerges on to the market.

It is interesting at this point to ask how well-established drugs would have fared under the FDA regime of testing. The first ever scientifically designed drug was salvarsan, which was discovered in 1909 by Paul Ehrlich as a cure for syphilis.[11] This disease results from sexual contact in which one partner passes on to the other a spirochaete, which is a corkscrew-shaped bacterium called *Treponema pallidum*. Ehrlich reasoned that it should be possible to find a poison which would kill the micro-organism circulating in a victim's blood stream without affecting the person. He thought an arsenic compound might be capable of doing this, and he started to make so-called arsenicals, testing them on rabbits infected with the disease. When he tested his 606th compound it worked, and salvarsan and hailed as a great triumph. The scourge of syphilis was lifted.[12]

The success of salvarsan proved what chemistry could do. And yet today it would not even have been submitted for testing because of

[11] The compound was actually prepared by Alfred Bertheim.

[12] Salvarsan was eventually superseded by penicillin. It is now held in reserve in case an antibiotic resistant strain of syphilis emerges. If this were to happen it is most unlikely that the syphilis bacterium would ever mutate in a way that would allow it to resist arsenic.

the arsenic it contains. And if salvarsan were submitted because of its effectiveness it would not to secure FDA approval because arsenic compounds are carcinogenic, and so it would fail the Delaney clause in the 1958 amendment to the US Food, Drug and Cosmetic Act. In effect, this simply states that no chemical which can be shown to cause cancer in human or animal can be approved for general use.

Other drugs that would fail today are the opiates, because they are addictive, and aspirin, because it causes internal bleeding. Needless to say, natural remedies, such as dried plants or seeds, which are sold as alternative medicines would never pass the tests demanded of chemically derived drugs. For a start they contain thousands of compounds, some of which are bound to be carcinogenic.

The biggest breakthrough in modern drugs came with penicillin, which is a fairly simple molecule. Thalidomide was discovered too soon, and that worked to its disadvantage; penicillin was discovered too soon and that worked to its advantage. It is produced by a mould, and it is deadly to certain bacteria. Had it not been for the exigencies of the Second World War, penicillin would never have been approved for human use since it would never have got past the testing stage—it has the disadvantage of killing guinea pigs (see the Box below). But penicillin was used and with remarkable results, saving millions of lives.

Since penicillin was such a success no one demanded that it be banned, even when its adverse side effects were noted. A few people are allergic to it, often violently so. Some have died from it. The millions of lives saved by penicillin have more than paid for these deaths, and yet the fear of being the unlucky one who has to pay that price has led some people to refuse to take any treatment that involves a chemical drug. It has been estimated that as many as 25% of all prescribed medicines are never taken by patients.

Guinea pigs

You may be pleased to know that these days guinea pigs are rarely used as guinea pigs. These unhappy creatures may have a lot in common with humans, in respect of their physiology and their inability to make vitamin C, but they tend to die rather easily. Dioxins kill them, and so does penicillin. It was their sensitivity to the former compounds that resulted in the dioxins being called the most dangerous chemicals known, as we shall see in Chapter 7. As little as a *millionth* of a gram is enough for a guinea pig. Guinea pigs are not much use when it comes to testing chemicals for their toxicity. Rats make better guinea pigs, and they have an in-built advantage: they cannot vomit. If you can persuade or force a rat to eat something which you suspect of being very toxic then you can be sure that the animal's body can not just spew it back at you. It has to digest it.

Hazard identification, by giving chemicals to animals, is the first step to determining their safety. If it does not kill them, then the next stage is to find out how much of a chemical can be tolerated. The final stages of testing may be the notorious LD_{50} trials, short for Lethal Dose 50%. The name refers to the amount necessary to kill 50% of a group of animals such as mice or rats (this test is now being phased out). If the chemical being tested has no cancer forming capabilities and low toxicity, in other words a high LD_{50} value, then it can proceed to tests on larger animals, and then on primates such as monkeys, before it finally goes for testing on humans.

Risk

If you want a treatment for a medical condition that is entirely without risk or side effects, then you must avoid all pharmaceutical preparations and surgery. Alternative medicine can offer you treatment that is almost risk-free, such as homeopathy, drinking or bathing in spa waters, aromatherapy or faith healing. However, these may not always be effective. How much risk you are prepared to accept depends on the severity of the disease your body is fighting. Sometimes the odds are so much in your favour that the fear is irrational, and yet the fear may be so great that it provokes a response out of all proportion to the danger, and people would prefer to suffer rather than take the prescribed treatment. Strangest of all is that the most opposition to a drug generally comes from those who have never been exposed to it. Which brings me back to the case of an NSAID that highlights the degree of risk we are talking about.

The real Opren scandal

Among the NSAIDs that emerged as a result of research by pharmaceutical companies was a painkiller that was given the name **Opren**. This worked better than aspirin and did not have the side-effects of causing internal bleeding. However, it did have side-effects of its own: of those treated with it a few people reported being very sensitive to sunburn, and a few old people died unexpectedly. Opren was blamed, and after a media scare the drug was taken off the market in the UK. Hundreds of thousands of people who were taking Opren were put on to other, less effective painkillers.

Of course the reason for withdrawing Opren was that it was less than perfect, and taking it did involve a degree of risk. What that risk was can be discussed numerically. If 25 000 people are operated on for arthritis, about 400 of them will die from the effects of that surgery, or from post-operative complications. If those 25 000 were to be prescribed Opren, so that they could live with their arthritis, then only *one* would die from side-effects of the drug. In gambling terms your odds of dying as a result of surgery are about 60 to 1, whereas with Opren treatment it was about 25 000 to 1. If you were to bet on which treatment would be most likely to be phased out you would have lost you money. Surgery continues, Opren was banned.

The statistics of life and death

We can imagine a drug with zero risk. It would target only the infecting organism or faulty cell, killing the former and repairing the latter. All healthy cells would be unaffected, and the amount of the drug that was surplus to requirements would simply be excreted by our body. In addition we would want the product to be safe enough to be sold over-the-counter, and to know that a child could take an accidental overdose and still be all right. One substance which meets these safety requirements is water, which explains why homeopathic medicines operate with zero risk. Water is the only active ingredient in such medication.

Most people opt for conventional medical treatment and have to live with the risk that attends all chemicals used as medicines. Many over-the-counter medicines are toxic when an overdose is taken, and some can have this effect on individuals even with a normal dose.

Many drugs have unpleasant side effects, such as the chemotherapy drugs used to fight cancer, and the risks are high. But what level of risk are we prepared to accept? By risk I refer, of course, to the risk of dying. This attends us every day of our life, and we generally accept the risk of a car accident, or climbing a ladder, and some of us even look for high risk adventures in leisure and sporting activities, such as climbing a mountain, hang gliding or underwater exploration.

These are clearly much more risky than taking medicine, and yet we will accept these risks provided we feel we are in control. We may even pass this control over to someone we trust, such as a pilot, or a surgeon. We are much less inclined to trust the tablets which a doctor prescribes, and we have become very suspicious of the chemists in white coats who make them. This puzzles chemists like me, and while we can quote statistics to show how safe our products really are, such messages do not seem to get through. The reason may be that our arguments appeal only to our rational brain, whereas our emotional brain is where the fear springs from.

However, there is a chance that if I can persuade you with logic, your emotions may follow. Suppose you need minor surgery for a condition that is not life-threatening, and the operation requires you to be given an anaesthetic. Once upon a time the risk of dying from the anaesthetic was as high as 60 in 10 000, but today it is less than 1 in 10 000. Would you go ahead? You probably would, even though it is difficult to place such a risk in the context of other rare events. But there is one event we all share—being born. The chance of your baby dying at birth, even in the most advanced countries of the world, is about 120 in 10 000. The chances of a woman dying while giving birth is 1 in 10 000, and this is also about the risk of having triplets. You can get a feel for a risk with such odds by asking yourself how many times you have personally known a woman who has died in labour, or who has given birth to a set of triplets. Like most people, you may know of neither.

These, then, are the baseline statistics of *risk*. The subject is dealt with in the award-winning paperback *Living with Risk* published by the British Medical Association Board of Science, which covers all aspects of risk such as smoking, alcohol, diet, occupation, sport, transport, and disease. Logically we should be prepared to accept odds of 1 in 10 000 against a medical treatment causing us to die.

The book reveals that we live with similar odds in other areas of life. For example, in the UK each year 1 in 10 000 of us will die in a car accident or a skiing accident, and 10 in 10 000 will die as a result of an accident at home. Many people are afraid of being struck by lightning, even though the chances of dying this way are about one in ten million (0.001 in 10 000). Which all goes to show how powerful our emotional brain really is. Yet of those who fear a bolt from the blue, how many are smokers? Of 10 000 smokers, 2000 will eventually die as a consequence of their addiction to nicotine.

According to statistics given in *Living with Risk*, people who work in the chemicals industry are safer at work than if they stay at home for the day. That may sound wrong, because every year we read of explosions and fires at chemical plants and the people there are working with some of the most dangerous materials known. Nevertheless it is true. Workers who use chemicals in industry and in agriculture are more likely to die from the beer they drink after work than from the chemicals to which they are exposed. Again that sounds wrong, but again it is true. People in restaurants, shops, offices and at home are exposed to solvents in cleaning fluids, pesticide residues, atmospheric pollution and lots of other chemicals. But they are more likely to be electrocuted than to die by inhaling such chemicals. Again you might think I am simply turning statistics on their head to shock you. I am— but only because these things are true.

Why then do we demand better odds on drugs than 1 in 10 000? For over-the-counter drugs the risks are probably less than one in a million, and we need give risk only a millisecond's thought as we take them. For more serious illnesses the chances must of course be greater. Anything which brings relief to human suffering will have side-effects for a few and be fatal for one or two. We must accept that this is so because there is no way we can change the chemistry of existing drugs, or the human body, to produce zero risk.

As our expectation of living a long life has increased, so we tend to forget the medical and pharmaceutical advances that make it possible. The healthier and longer our life, the more obsessed we have become that it should be even healthier and even longer. Books, magazines, newspapers and television are full of news and advice about things that will help us achieve this, but our biggest asset is the human body itself. It is much better equipped than we realise to protect us against all sorts of threats, since it has evolved to deal with many of them. Every cell of our body probably undergoes some malfunction every week of our life, and each time a wonderful battery of natural chemicals rushes to repair it. So long as we do not abuse it and so long as we protect it from infectious diseases, then it will serve us well.

Our fear of the side-effects of drugs is irrational, and this has undermined our faith in them, predisposing us to go along with campaigns to reject new drugs at the first sign of any risk. Yet if we are to make progress against the illnesses which cut life short we must continue to support research in this area. There are still many conditions which make life miserable for young and old. If we want a cure for acne, arthritis or AIDS we must go on. The answer to these afflictions can only come from the test tube, whether it is discovering new drugs or extracting hitherto unsuspected drugs from natural sources, and developing safer variants of them. We can even find a cure for cancer—but only if we are prepared to take the risk.

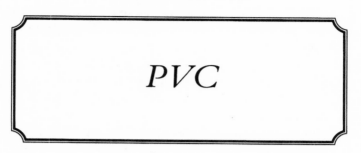

CHAPTER SIX

PVC

*In praise of plastics; the manufacture of PVC; the additives
and hazards of PVC; the environmental audit*

THE plastic we have learned to avoid is **PVC**, which is the abbreviation for **poly(vinyl chloride)**.[1] There appear to be three good reasons for this: PVC is made from a liquid which causes cancer; when it is used as packaging it contaminates our food; and when it burns it forms dioxins. All these things are true, yet the risks they pose to our health are so minuscule that they can be ignored. Which is why PVC is still used to make bottles for mineral waters, and containers for blood. It may come as a surprise to learn that PVC is the plastic chosen for catheters, the very fine tubes that are inserted by doctors and surgeons into the human body, even into premature babies. Nor need special care be taken to sort out PVC from other domestic waste before this is burnt in municipal incinerators, because it will produce no more in the way of dangerous dioxins than other materials that are burnt.

Not everyone sees the widespread use and benefits of PVC as tacit approval by society that this plastic is safe. Some environmentalist groups, such as Greenpeace, would like to see PVC banned altogether because it is an organochlorine compound.[2] As a result of

[1] A reminder that when the name of a substance appears in **bold** type, it means that there is more about its chemistry in the Appendix.

[2] An organochlorine compound is defined as one that has a carbon-to-chlorine bond in its molecule.

their alarms you can sometimes see the words 'not made from PVC' printed on a product, as if those that were made from PVC should carry a health warning. However, it is going to be very difficult to replace PVC because this plastic offers many benefits. It is a superb electrical insulator, it is flexible, it is tough, it is scratch-resistant, it does not crack, and it does not catch fire easily. PVC products last for decades and they are cheap to make.

Because PVC resists corrosion and weathering it is used for electrical wiring, water pipes, water butts, window frames, mud flaps, floor covering, fences and garden furniture. It recycles well and some of these outdoor products are indeed made from recycled PVC. Another benefit of this plastic's wide use is the employment it gives to the 70 000 people around the world who make it. Or should we really think of them as victims of an industry which puts them at risk by exposing them to undesirable chemicals? Indeed if what is said about PVC is true then we may all be at risk. However, before I discuss the health and environmental aspects of its use, I would like to set it in the context of other plastics.

In praise of plastics

PVC is a plastic, although it is more correct to call it a polymer.[3] Chemists were once greatly admired for the new materials they kept discovering, not least of which were a series of wonderful plastics. These replaced the older materials that were derived from animal and vegetable sources, such as leather, latex, cotton, ivory and wood, materials we would now call biopolymers, signifying that they are derived from living things. Nature finds such polymers indispensable, of course, but for our purposes they needed to be modified before they were suitable for such things as footwear, rubber, clothes, combs and books. But there were limits to these natural polymers, not only to what the materials themselves could do, but also to the amount which Nature could provide.

---

[3] Plastics and resins are the older words that came into use before we understood what these materials really were. The modern equivalents are polymers and cross-linked polymers.

Biopolymers have admirable qualities such as lightness, toughness and durability. Nature in its bounty evolved many natural polymers, such as strong wood, insulating wool and waterproof skin, but it tended not to give priority to features such as electrical insulation, high temperature stability, transparency and colour, all of which are regarded as important in manufactured products. Some polymers made from natural products have these properties: for example rubber is an insulator, it can be coloured, and early electrical wiring was sheathed with it, but a major drawback is that rubber oxidises slowly in air and after a few years becomes brittle and develops cracks.

Nevertheless many biopolymers are very good and perform very well, including fur, leather, lambswool, silk, mahogany, ebony and ivory.[4] The Industrial Revolution changed manufacturing methods and brought newer and cheaper products to an expanding population. To begin with the new industries simply produced more of the known materials more efficiently, but the vagaries of weather and disease could seriously interfere with supplies from distant countries, not to mention the other socially disruptive activities that humans are prone to indulge in: financial speculation, politics and war. These, and the growing pressure of an expanding population, encouraged industry to use the alternative materials that chemists discovered. To begin with, such products were welcomed by everyone.

By the second half of the nineteenth century the demand for ivory had grown to such an extent that 100 000 elephants were being slaughtered every year for their tusks. These consist of the peptide biopolymer keratin, and ivory was ideal for making combs, piano keys, buttons, cutlery handles, and billiard balls. These could also be made from the first synthetic plastic, which was **cellulose nitrate**. Alexander Parkes discovered this and it was put on display in the Great Exhibition in London in 1862. It was not completely synthetic because it was made from wood cellulose, but he modified this chemically rather than merely processing it. Cellulose nitrate, or Parkesine as it was called, was used to make many of the items that had previously been made from ivory. In the US an inventor John Wesley Hyatt was also experimenting with cellulose nitrate for billiard

---

[4] The dilemma is that these could be the materials of the future, because they come from renewable resources.

balls, but these were far from ideal. Sometimes they exploded when they collided violently, and one burst into flames when it came into contact with a lighted cigar. Cellulose nitrate is highly flammable and products made from it were always potentially dangerous.

All was not lost, because in 1869 Hyatt showed that cellulose nitrate could be turned into a better plastic by mixing it with **camphor**, and heating it under pressure. He called the product celluloid, and the first things he made from it were false teeth. He set up the Celluloid Manufacturing Company of Newark, New Jersey, in 1872 to produce a range of household goods, and these found a ready market. The firm continued in business for over 75 years until the factory closed in 1949.

The next big breakthrough, this time in artificial fibres, was also a modification of cellulose. When this is converted chemically into its acetate derivative, **cellulose acetate**, it can be drawn into fibres and woven to form artificial silk. This was first marketed about a century ago, and at a time when the natural fibre could no longer keep pace with demand. Cellulose acetate polymer, derived from wood pulp, is still made on a large scale today and called by a variety of trade names as well as the generic names of rayon, viscose, acetate and art silk.[5] This polymer is ideal for soft furnishing, carpets, linings and blouses, but it does crease easily. Another useful form of cellulose acetate was Cellophane, again a product that was greeted with enthusiasm when chemists finally mastered its production in the 1930s.[6]

Research chemists are still working to improve cellulose acetate, and a recent development has been a manufacturing method which keeps the long chain cellulose polymers of the wood intact, rather than breaking them down as happens with the older processes. The improved rayon is achieved through the use of a new type of solvent, which is also recycled in a process that reclaims 99% of it each time it goes through the process. The new rayon is produced by the chemical company Courtaulds and is called Tencel, and has features that rival the purely synthetic fibres such as **nylon**. It is not only very strong and soft to the touch, but it also takes dyes very well.

[5] The 'art' stands for artificial, although many think this is another form of real silk.
[6] Cellophane is still manufactured, but it is now in competition with polypropylene which is stronger.

Biopolymers are based on compounds of carbon that are made naturally by plants and animals. Synthetic polymers are also based on carbon compounds, and when they were first discovered the source from which they came was coal, a fossil fuel resource which seemed inexhaustible and of which there are still reserves of a trillion (10^{12}) tonnes. Supplies were also less prone to disruption. Chemicals from coal led to the first truly synthetic polymer, **Bakelite**, which was first made in the USA in 1909 and took its name from its discover Leo Baekeland. It was much safer than celluloid, and tougher, and it replaced this plastic in many of its applications. Bakelite was a thermo-setting resin[7] made from **phenol** and **formaldehyde**. It was stiff, light, reasonably strong, had insulating properties and could be moulded easily. It was the ideal plastic for electrical appliances, and it was Bakelite which made possible the generation and distribution of electricity, and made electrical appliances safer for use in the home.

Bakelite was not without its drawbacks. It came only in dull colours, and though it was a good electrical insulator it was a rigid plastic and useless for electrical wiring. Nevertheless, it made synthetic plastics acceptable and so paved the way for others that were to be discovered in the years between the two World Wars. This was the time when the chemical nature of polymers was finally explained by the German chemist, Hermann Staudinger. He referred to them as macromolecules (meaning large molecules) rather than polymers, and his theory was that they were built from smaller units that had joined together to form long chains. He was to be proved right.

By the end of the 1930s several purely synthetic polymers were in commercial production. Some caught the public imagination imme-diately. Stockings made from nylon, the polymer which had been discovered by the chemist Wallace Carothers in 1935, almost caused riots among shoppers when they went on sale in New York in 1939.[8] Other polymers that had been discovered about this time were **poly-ethylene** (polythene), **polystyrene**, **poly(methyl methacrylate)** (Perspex), and poly(vinyl chloride) (PVC). All were to be developed further during the Second World War. With the end of the war in

[7] Such a material is originally soft, but on heating it changes permanently to a hard rigid form.
[8] Carothers did not live to see the success of his polymer. He had committed suicide in 1937.

1945, the chemicals industry that was producing these plastics found a public eager to buy products made from them. Within a few years they were to become part of everyday living, and their names entered the language. By 1990 the yearly output of synthetic polymers reached 80 million tonnes a year, and all are derived from oil or natural gas.

Perhaps we have become so used to plastics that we no longer notice or appreciate the benefits they bring. Plastics do not rust, corrode, or break easily, and they can be permanently coloured. In many applications they are better than metals, and this has become particularly important in aircraft and car components where their lightness also improves fuel efficiency. Polymers can be turned into packaging, such as transparent film, rigid containers or unbreakable bottles, and again their lightness saves on transport costs. They have revolutionised building and furnishings, because unlike wood they do not rot, unlike wool they do not stain or wear, and unlike iron they do not rust or need painting. The modern electrical and electronics industries would be unthinkable without plastics. In hospitals they are used for disposable syringes, intravenous blood tubing and bags, all of which can be destroyed by incineration after use.

There are various classes of polymer depending on the type of molecule from which they are constructed. PVC belongs to the ethylene family, which has a backbone chain made entirely of carbon atoms, each with two hydrogen atoms attached. **Ethylene** gas, whose two carbon atoms are joined by a double bond, gives rise to the simplest polymer, polyethylene, which consists of chains of $-CH_2-$ units. It forms this by using one of its double bonds to link to another ethylene molecule, which then links to another ethylene molecule, and so on *ad infinitum*, or at least until there are no more molecules to be linked up.

In this way very long chains of carbon atoms are produced, and these have the wonderful properties that we associate with polymers. Derivatives of ethylene, in which there are atoms other than hydrogen attached to the carbon chain, also undergo the same type of reaction to give some very interesting varieties of polymers. These are listed in the table (next page). Among this list are the world's leading polymers: PVC, polyethylene and polystyrene. PVC consists of chains of carbon atoms with every other carbon atom carrying a chlorine, like beads on

Ethylene based polymers

Monomer*	Polymer	Uses
acrylonitrile	polyacrylonitrile (PAN), Orlon, Dralon, Courtelle	Knitwear, high impact plastics
ethylene	polyethylene, polythene	Kitchen utensils, containers, plastic bags, wire insulation, toys, etc.
methyl methacrylate	poly(methyl methacrylate), Perspex, Plexiglas, Lucite	Transparent objects, transparent panels, illuminated signs
styrene	polystyrene	Foam, lightweight cartons, insulation, etc.
tetrafluoroethylene	poly(tetrafluoroethylene) (PTFE), Teflon	Insulation, non-stick surface coatings
vinyl acetate	poly(vinyl acetate) (PVA)	Paints
vinyl chloride	poly(vinyl choride) (PVC)	Rainwear, window frames, pipes, adhesives, cling film (plastic wrap), bottles, flooring, insulation, furniture, car components, etc.

*These names are the older chemical names. There are also systematic chemical names for these compounds: for example, vinyl chloride is more correctly known as chloroethene (see the Appendix).

a necklace. PVC is used to make hundreds of products, and in some situations it offers just the right combination of properties. For instance, it is ideal for the colour–coded, flexible, long–life plastic coating for electrical wires. It also has another benefit: it is one of the cheapest polymers to make.

The manufacture of PVC

There are three steps in making PVC: first, the production of the monomer, vinyl chloride; then the polymerisation of this to form the polymer; and finally the blending and tailoring of the polymer to give the product we know as PVC. At all stages there are risks to be considered. In the first step **chlorine** gas is reacted with ethylene gas. Both these chemicals are hazardous: chlorine is toxic and ethylene is flammable. The product of their reaction is ethylene dichloride and then vinyl chloride, which is also hazardous and has caused cancer.

Ethylene comes from natural gas or oil, and is made by heating **ethane**, **propane** or **butane**, or it can be produced from naphtha

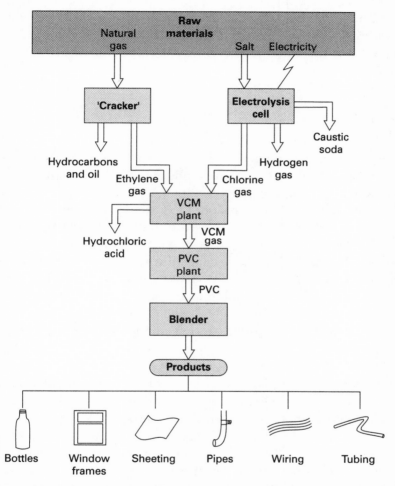

The production of PVC. This shows the main stages in converting salt and natural gas into hundreds of manufactured articles that are made from this plastic. Each stage consists of several steps, including reclaiming unused material and recycling it through the system again. In addition there are by-products of the various processes such as caustic soda, hydrocarbons such as propylene, and hydrochloric acid.

which comes from oil. **Methane** and hydrogen are given off as by-products during the process, but these gases can be burned to provide energy on site. Propylene is also produced, and this is valuable enough to be reclaimed. Chemists are familiar with these materials, and though they are flammable they are neither toxic nor cancer forming. It is the chlorine and vinyl chloride about which there has been most concern.

Chlorine

Because chlorine is needed to make it, PVC now finds itself at a disadvantage. Chlorine gas is produced from sodium chloride (common salt), and the manufacture of PVC accounts for about 30% of the chlorine that is produced industrially. Chlorine is a highly corrosive and lethal gas that was used in the First World War, to kill thousands of troops.[9] Today it is used to make bleach and purify water supplies. Of course chlorine is dangerous to handle, and there have been industrial accidents involving it in which people have died. Official figures show that chlorine has killed a total of 81 people throughout the world since 1960, which is about three per year on average. At the time of writing, the last fatal accident in Europe was at Málaga in Spain in 1974, where the gas was being used to purify water supplies.

Most chlorine is produced at the site where it is to be used. Some has to be shipped to where it is needed, and indeed millions of tonnes of *liquid* chlorine are moved by road, rail and sea each year. It is stored and transported as a liquid under pressure because in this form it takes up much less room. When it is destined for industrial use it is transported in steel containers in quantities of up to 50 tons. For non-industrial uses, such as water sterilisation, it is delivered in 50 kg containers. All this chlorine on the move sounds very frightening, and there are occasional accidents, but none of these has yet compared with the spectacular fireballs that have engulfed motorists when petrol tankers have crashed and exploded on motorways. Hopefully

---

[9] It was eventually superseded by much more toxic vapours, such as mustard gas and lewisite.

the margins of safety surrounding chlorine are stringent enough to ensure that a major spill will never happen.

Official records are reassuring, and show that the precautions and regulations which govern chlorine are capable of reducing the number of accidents to a minimum and of dealing effectively with those that do occur. In the 40 years from 1950 to 1990 there were 90 major incidents involving road and rail tankers transporting chlorine. In these a total of 22 containers were ruptured and 36 tonnes of chlorine escaped. Luckily no one died. There were also 32 accidents involving small containers of chlorine of which 25 were broken with the loss of 10 tonnes of chlorine. Four people died, all in the one incident in Málaga mentioned earlier.

Where does chlorine come from and how is it made? There are unlimited supplies of salt in the sea, but more accessible are the large deposits which are the remains of dried-up prehistoric seas, and which are to be found in many countries. When salt is dissolved in water it forms a solution known as brine, and if a current of electricity is passed through such a solution in a special cell, it releases the elements of sodium chloride: chlorine gas bubbles off in one part of the cell while sodium metal is produced in the other.[10] The sodium then reacts with water to form sodium hydroxide (**caustic soda**), at the same time releasing hydrogen gas. Both of these products also have their own commercial uses.

The equipment used to generate chlorine gas involves the liquid metal mercury, and this causes environmental problems because its compounds are toxic. Over the years there have been several escapes of mercury, sometimes with disastrous consequences, such as that which happened in the 1950s in Japan at Minamata Bay. There the mercury contaminated fish and so entered the food chain, and as a result local people were poisoned. Some died, others suffered brain damage. Today, a mercury disaster is looming in Brazil where miners use it to extract gold from the sediment of rivers. For every ounce of gold produced, several ounces of mercury are lost.

In theory there should be no loss of mercury in the process to generate chlorine, but in older plants there was substantial loss of this toxic metal to the environment where it polluted water and

---

[10] See the Appendix under chlorine for a description of a chlorine cell.

soil. This was rightly seen as unacceptable, and such have been the improvements in cutting the loss of mercury, that in some plants it may be as little as a quarter of a gram for each *tonne* of chlorine produced. However carefully the process is carried out, there is always some loss of mercury, and this is why this method of manufacturing chlorine is on the decline. For example, manufacturers of chlorine gas around the North Sea are pledged to phase out the mercury process by the year 2010.

Chemists have devised two other ways for making chlorine, and neither involves mercury. In the so-called diaphragm process there are two compartments in the cell, one producing chlorine and one producing caustic soda and hydrogen, and the brine in the two compartments is kept separate by an **asbestos** diaphragm which still allows electric current to flow. Asbestos is used because this is porous and resistant to the corrosive action of both chlorine and caustic soda. You may think that this has merely swopped one hazard for another—mercury for asbestos—but there is no loss of asbestos during the process, and when the diaphragm needs replacing it can be disposed of by burying it safely on site.

The 'cleanest' technology for producing chlorine is the membrane process. Here the two compartments are kept separate by a membrane made from a polymer material which is highly resistant to chemical attack. The method is more expensive to run because it leads to a more dilute solution of the caustic soda which has to be concentrated to make it saleable. To do this the excess water has to be removed by steam heating, which takes a lot of energy, and even then the final product is contaminated with brine.

Vinyl chloride

When ethylene and chlorine gases react they form a liquid, **ethylene dichloride**. This was once widely used as a dry-cleaning solvent, but has been phased out on health grounds because long-term exposure to its vapour was damaging to the kidneys. Ethylene dichloride is then turned into vinyl chloride by heating, which causes it to lose hydrogen chloride gas. Things are bit more complicated than this, and side reactions also occur which lead to various other organochlorine compounds, including minute traces of dioxins. Some of the by-products are worth collecting because they have a commercial market, although this is declining as the damand for organochlorines declines. Those that are not wanted can be burned to reclaim their hydrogen chloride, which can be recycled and reacted with more ethylene to make another batch of ethylene dichloride.

Vinyl chloride is a gas which boils at -14 °C. It is harmless compared to chlorine, but it is not without its drawbacks. Vinyl chloride was even tested as a general anaesthetic in the 1940s, and until the 1960s it was regarded as perfectly safe, provided it was treated with respect because it is potentially a flammable gas. The first inkling that all was not well came from those working with it. They reported a numbness in the tips of their fingers. Was vinyl chloride to blame? Tests were carried out on rats which were exposed to high concentrations of the gas. These tests did not reveal the cause of the numb fingers, but they did reveal something far more worrying—a rare form of liver cancer called angiosarcoma. More tests were ordered and the results were confirmed. A study among the health records of all who had worked with vinyl chloride in the USA found cases of people with this type of cancer.

Government health agencies, trade unions and managers in the industry were alarmed, and cooperated to bring about major changes. Leaks and losses of vinyl chloride gas in plants were reduced to the absolute minimum; workers exposed to it were to be protected; any residual traces of vinyl chloride in the product PVC were to be removed as far as possible. These changes are now part of national and international law wherever PVC is made. Today there are no cases of numb fingers caused by exposure to excess vinyl chloride, but there are still a few new cases of angiosarcoma being reported. The reason for this is that it takes about 25 years before someone heavily exposed to vinyl chloride contracts the disease, so that people exposed in the 1960s and early 1970s may still develop angiosarcoma. It will be a few years yet before we have seen the last of this rare industrial illness.

The first case reported was in 1964 at a plant in the USA. Of the 1800 people who worked at that plant, only ten have so far contracted the disease. By 1988 there were 200 known cases of the disease worldwide. Levels of vinyl chloride to which workers are exposed today are kept below 1 part per million (ppm) of the air wherever possible. In the cancer tests on rats they breathed air in which there was 30 000 ppm for 20 hours a week for a year.

Might cases of angiosarcoma yet appear among the general public? This is almost certainly not going to happen. Angiosarcoma is a very rare complaint, and no member of the public has ever been identified as contracting this disease because of exposure to vinyl chloride. Of course when polymerisation of vinyl chloride takes place there is always a little of this material trapped in the final PVC. Improvements in technique and chemical know-how can keep this to a minimum, but they cannot exclude it totally. Incomplete polymerisation during manufacture once left a significant amount of vinyl chloride in the finished PVC. Today's regulations demand that if the PVC is destined for food containers then the level of vinyl chloride must not exceed 1 ppm, and most of this evaporates during subsequent processing of the plastic. For general purposes 5–10 ppm of vinyl chloride is the limit, but again much lower levels are to be found in the final product.

Even so, you may wonder if the tiny trace of vinyl chloride could escape from articles made of PVC and affect those working with it, or even consumers who are using products made from it. Could some of us get cancer because of it? The short answer is no. The most likely way in which you would be exposed to the vinyl chloride residues in PVC is by breathing the dust if you work with it in industry. PVC dust can irritate the lungs just as any dust does, but there is no evidence that it causes cancer, and in any case the working environment for PVC dust is set at the same low level as for all such dusts. Nor does heating PVC cause it to revert to vinyl chloride. The general public is at no risk at all from this chemical. You can stop worrying about catching cancer from contact with PVC—it never happens.

What about people who live in towns where PVC is manufactured? They are not protected by special safety gear like those working in an industrial plant. Are they at risk from vinyl chloride? Investigators in Sweden and Norway compared the health of those living in communities where there were PVC plants with the health of people living in towns where there were no PVC-based industries. They found no difference in the incidence of disease. This is reassuring, and shows that the regulations now in force are working. These regulations are designed to prevent any obnoxious material escaping from such plants into the environment. Of course tiny amounts of all the materials used in making PVC do escape—that is unavoidable—but detecting these is rather a tribute to the skills of public analysts and their monitoring equipment. We shall return to this aspect of analytical chemistry in Chapter 8.

Under pressure vinyl chloride becomes a liquid, and it is in this state that it is used to make PVC. The process is carried out inside high pressure chambers. The reaction is usually performed with the vinyl chloride dispersed in water, as a suspension or an emulsion. The reaction in water is done at temperatures of 50–70 °C, and the water controls the process by removing the heat that is given off during the polymerisation. The PVC appears as tiny particles, and when these have grown to the right size the reaction is stopped and any vinyl chloride that has not polymerised is distilled off and recycled. The PVC is separated and dried in a stream of hot air.

The additives and hazards of PVC

Are there any health hazards to the public in using PVC? Again the simple answer is no. Yet people are worried about PVC and it continues to receive adverse publicity. One worry is the trace of vinyl chloride, which we have seen is now so low it can do no harm. The second worry concerns incineration of PVC waste and whether this releases dioxins. There is also a small emission of dioxins from the plants which make PVC. The third worry is the other components which are put into PVC to make it a suitable plastic for the job intended. Are these additives safe?

Let us look at the more dangerous toxin first, dioxins. The disposal of PVC could give rise to dioxins if it were burned. These chemicals are given off when household waste is incinerated, and for a time it looked as though the main source of dioxins might be the PVC waste. PVC has all the right elements for making dioxins, but the amount of dioxins given off from burning PVC is tiny. This was graphically demonstrated in Homsund, Sweden, when a wooden warehouse caught fire. Inside it were stored 200 tonnes of PVC and 500 tonnes of plastic carpets, all of which were consumed in the conflagration, which produced a cloud of smoke. Due to some rather unusual weather conditions at the time—the temperature outside was -30 °C—the cloud of smoke from the fire was trapped by atmospheric inversion and most of it came down on to fresh snow within a radius of about a mile of the warehouse[11]. Samples were taken and analysed for dioxins. As we might have expected they were detected, but in surprisingly tiny amounts. The investigators calculated that only 3 milligrams had been produced in the fire, and most of that had probably come from the wooden building itself. We shall return to this in Chapter 8 when I will discuss dioxins in more detail.

So is there nothing really to worry about burning PVC? Well, not quite. It is common practice to reclaim copper and steel wire by burning off the PVC coating. This is a dangerous thing to do especially when copper wire is involved. This metal has the chemical property of catalysing the formation of dioxins from PVC fumes. A better method is to strip the PVC off mechanically and recycle it separately.

Public concern about PVC did not arise from the vinyl chloride residues or from the dangers of dioxins. The alarm came over the

The modifiers (additives) used in PVC

Modifier	Average (%) in PVC when used*
Common modifiers	
Plasticisers	30
Fillers	30
Processing aids	5
Colours	2
Heat stabilisers	2
Lubricants	0.5
Specialist modifiers	
Fire retardants	4
Antistatics	1
Flow modifiers	1
Foaming agents	1
Biostabilisers	1

*Not all additives are used in all types of PVC—these are averages of quite wide-ranging values. For example, plasticisers may be as little as 10% or as much as 50%.

[11] In this unusual weather pattern there is a layer of warm air trapped below a layer of heavy cold air.

additives which PVC contained. The simple polymer produced from vinyl chloride has to be blended with up to ten other materials to create the texture, colour and stability of PVC we desire. These additives may account for a large part of the finished product, as the table opposite shows. Indeed, in some cases they account for most of it.

Plasticisers are the key to the versatility of PVC, which by itself is a rigid material. They are needed to make it flexible, and how flexible it is depends on how much plasticiser is added. The compounds used for this purpose include natural oils and other polymers, but the best plasticisers are **adipates** and **phthalates**. Half the weight of the finished product may be plasticiser, and for very flexible PVC, like the transparent film used to wrap food, the plasticiser has sometimes proved a problem. When it is in contact with fatty foods, such as certain types of cheese, the plasticiser can migrate from the plastic film to the product. The realisation that this was happening led to a public outcry, and although such film was very useful in the kitchen in keeping food clean and fresh, many people stopped using it because they feared that the plasticisers were capable of causing cancer.

The other additives in PVC are designed to make it hard or soft, insulating or conducting, transparent or opaque, or to colour it. Some of these additives may slowly work their way to the surface and be lost during the lifetime of the article. In theory all the additives are safe for the application for which the plastic is intended, and where the PVC comes in contact with food, or human tissue, the additives employed have passed recognised safety tests. Lead compounds cannot be used as additives for food containers, but they are allowed for insulation on underground cables. Phthalates cannot be used in fridges, but they are permitted for carpet edging. So what are additives designed to do? Could some of them really be health hazards?

Plasticisers   To make PVC flexible we need to add chemicals that will allow the chains of polymer to move past one another smoothly. Up to half the weight of the plastic may be plasticiser, and this may become a problem if the plasticiser migrates from the PVC to something that is in contact with. Phthalates, adipates and organophosphates are used as plasticisers. **DEHP**, short for di(2-ethylhexyl) phthalate, accounts for more than half of all plasticisers put into PVC. Another common plasticiser is **DEHA**, short for di(2-ethylhexyl) adipate. DEHP is used in PVC which comes into contact with food, and it is the only one permitted for medical uses such as blood storage bags, catheters and other devices. DEHA is approved for use in cling-film and food containers. Both DEHP and DEHA are non-toxic, and you would probably survive even if you ate two kilograms of either. Their long term safety was put in doubt in the 1980s when it was shown that mice and rats fed DEHP had high incidences of cancer. Whether these compounds cause cancer in humans is not known (see the Box, below) but since people only come into contact with minute amounts their use probably carries no risk.

The American National Cancer Research Institute carried out tests with rats and mice, giving them an amount of DEHP which was the equivalent of two coffee mugs full every day for life for a human. Not surprisingly, given these massive doses, the rats developed cancer. The International Agency for Research on Cancer said there was a risk, but the European Union Scientific Committee on Food says they can still be used, and has set an acceptable maximum daily intake of 25 mg/kg for DEHP and 30 mg/kg for DEHA. This corresponds to an intake of 1750 mg (DEHP) and 2100 mg (DEHA) per day for an average adult weighing 70 kg. Nothing approaching this amount would ever migrate from a piece of PVC.

Cancer testing

If we suspect that a chemical will trigger cancer, how do we test it? Clearly we must test it on animals, but the more these resemble humans, the harder and costlier it will be to carry out the test. Indeed, even to expose primates like monkeys and chimpanzees to such tests is almost unthinkable, let alone human volunteers, however willing they may be. Testing rats and mice is more acceptable since we normally think of these animals as rodents and vermin, although of course those reared for testing are rather special creatures.

In 1984 David Haseman and Hans Zeisel discussed the validity of testing mice and basing predictions about the effects on human beings from the results. In one piece of research 88 chemicals suspected of causing cancer were tested on both rats and mice. Of these, 17 compounds caused cancers in both animals, 14 in mice but not rats, and 12 in rats but not mice. Six compounds that were known to start cancers in both groups of rodents were then tested on monkeys, but only one of them was found to cause cancer in these animals.

It may be true that what causes cancer in humans also causes cancers in rodents, but the reverse does not necessarily apply. When next you hear of some food component, or environmental contaminant, causing higher-than-expected cancers in mice and rats, do not automatically assume that it will do so in humans. The chances are it probably won't.

Heat stabilisers PVC would decompose when heated if it were not for heat stabilisers. These are metal salts or metal soaps.[12] Tin, calcium, barium and zinc salts are the metal compounds most commonly used. Lead salts are also used for such things as electrical cables, pipes and window frames. Tin salts stabilise the clear rigid PVC of bottles. Could these metals taint the contents? The answer is no, because the amounts which migrate from the surface of the plastic are insignificant. Lead-stabilised PVC water pipes are in use and are acceptable in Europe, although the USA has now gone over to tin-stabilised pipes. Once the lead-containing pipes have been installed they are thoroughly flushed to remove the lead additives at the surface, after which they perform well within acceptable limits.

There is growing concern that no heavy-metal compounds should come into contact with food, toys and medical devices, and we can expect to see such stabilisers phased out. Another reason for phasing them out is that they contribute to the burden of heavy metals in the environment, especially when they are disposed of. The metals will be released extremely slowly if the PVC is disposed into a landfill site, but very rapidly if the plastic is incinerated. PVC is highly resistant to biodegradation, and so far there is no evidence that heavy metal stabilisers are released from discarded PVC. On the other hand flue-dust and slag from incinerators is assumed to be toxic waste and treated as such. It is taken to special landfill sites.

There are some heat stabilisers which do not use metals. They are made from unsaturated natural oils such as soya bean oil, sunflower oil

---

[12] These are the salts of fatty acids.

and linseed oil. When this type of PVC is incinerated, the organic stabilisers simply end up as carbon dioxide. Nevertheless some of these organic stabilisers can not be used by themselves, and there has to be some metal soap put with them. For example the PVC for medical applications, which uses mainly organic stabilisers, has to have a combination of zinc and copper soaps as well.

Lubricants   These are added to PVC to control its rate of flow when it is molten, and to prevent it sticking to metal surfaces, especially to machinery. Lubricants, which are organic waxes or organic acids, never amount to more than 1% of the polymer, and sometimes only 0.1% is needed.

Processing aids and impact modifiers   These may make up 10% of the finished PVC, and they consist of other polymers. Their purpose is to make the plastic more workable or make the final article stronger. If the final product is to be used for food containers or medical applications, then processing aids and modifiers which are used have to meet approved specifications.

Fillers   These may contribute up to half the weight of a PVC plastic article. They are there because they are cheaper than the PVC they are displacing, and so they increase profits. Powdered chalk or limestone are the commonest fillers, and they pose no health or safety risks, either when the polymer is in use or when the PVC is disposed of.

Colourants   These are either organic dyes or inorganic pigments, and care has to be taken if the intended use of the PVC is for food containers or medical equipment. Traditional pigments such as lead or cadmium compounds may seem to be safely locked away within the structure of the plastic, but they are intrinsically toxic and they can be released, for example if a child chews a piece of plastic and swallows it. Such colourants are no longer acceptable in either toys or paints, and not only because of the risk to children. Workers who manufacture or dispose of such materials are also at risk of exposure, and these colourants are now controlled by various regulations. Organic dyestuffs may appear less of a threat, but again these must be tested for safety because in some applications, such as high temperature processing of materials, they may give off vapours which are more toxic than the fumes of metal-based pigments.

Flame retardants   PVC is less flammable than most polymers, but it can be made even safer by adding a few per cent of a metal oxide such as **antimony oxide**. A little added molybdenum oxide also reduces the amount of smoke given off when the plastic burns. Here there is an environmental price to pay for these safety benefits, because these metals are toxic, but the benefits of flame retardants still outweigh their disadvantages. Rigid plastic objects, however, need no flame retardants.

Other additives   So-called blowing agents are needed to produce a foam texture in PVC by creating gases like carbon dioxide or nitrogen which expand the molten plastic with lots of fine bubbles. Other specialist additives are biostabilisers, antistatic agents and viscosity modifiers. As with all the major additives they have to be tested for toxicological safety and environmental compatibility before they can be used.

The environmental audit

Making PVC depletes natural resources, manufacturing it produces dangerous chemicals, and disposing of it adds to the environmental burden. The same is true of every synthetic plastic, indeed it is true of almost all manufactured materials that humans have used throughout recorded history, including such traditional products as iron, lead and glass. People have always been aware of the threat these activities posed to their local environment, but nowadays the global environment has to be taken into account. Today it strikes people as folly that our grandparents regarded the Earth as an exploitable resource. Minerals were mined regardless of their location, the desolation it caused, and the disruption to the habitats of other species. Solid waste was piled up in slag heaps, liquid waste was pumped into rivers or dumped in the sea, and toxic fumes were vented to the air. Domestic rubbish was tipped into natural or man-made holes in the landscape and covered with soil. Out of sight, out of mind was the general attitude, and Nature was left to look after itself. Indeed, industrial waste seemed a necessary evil that came with flourishing industry, full employment, fat profits and cheaper products. The ends justified the means.

Thankfully attitudes have changed, and the 1960s saw the start of a public debate about pollution and the environment which has continued to the present day. Talk gave way to action as people sought ways to reduce the mountains of waste that everyday modern living seemed to be generating. In fact industry, mining and agriculture are still the major producers of waste, but for most people it is the waste they generate themselves that they feel they should do something about. Packaging, and particularly plastic packaging, falls into this category, and yet plastics make up less than 10% of all refuse, and PVC only about 1% of all waste. Paper, glass and metal still account for most of what we throw away.

Against this background, how then do we assess the environmental cost of something like PVC? Ideally we should carry out a cradle-to-grave analysis for both the synthetic plastic and the traditional materials it replaces. One day this will be done, and already there are techniques being developed to carry it out under such titles as Lifecycle Analysis and Ecoprofiling. These methods of analysis divide the problem into conservation, which deals with the production and distribution of a product in terms of renewable and non-renewable resources, and pollution, which covers the environmental costs after the article has ended its useful life. The longer its working life and the safer it is, the better. We might expect PVC to come out badly in such an audit because it is the least loved of any plastic. We lack the data as yet to do a full environmental audit for PVC, or any biopolymer for that matter, but we can do a partial audit.

Despite strict controls and guidelines that have made PVC safe for food packaging and medical use, people still remember the time when it was considered dangerous. Some people consider it an environmental hazard, and it finds itself grouped with other organochlorines such as **DDT**, **PCB**s, **CFC**s, dangerous solvents and dioxins. In fact PVC poses little threat to the environment, and is no worse than other synthetic polymers and biopolymers; and there are uses which demand PVC for reasons of safety and economy.

Many materials carry a high environmental price, very high indeed if they threaten a species with extinction by destroying its natural habitat. Even crops such as those used for fibres and paper, which we think of as benign and of which we may approve, still require the using up of natural resources: they require fertilizers to maximise growth; fossil fuels for harvesting and transportation; and energy and chemicals for processing into the final products. PVC cannot be accused of depleting natural resources to any greater extent than traditional materials. To make PVC requires rock salt, of which supplies are virtually unlimited, and hydrocarbons, which are ultimately of limited extent but are at present very abundant. Biopolymers get their carbon from the carbon dioxide of the air and in this respect they score over PVC. In theory PVC could also be made from renewable sources—salt can be extracted from seawater, and ethylene from crops—and one day it may have to be so made, if we are to continue to enjoy the benefits of this versatile plastic. Then it will cost more to produce, but so will all plastics.

Pollution The part of the environmental audit where PVC appears to build up a large deficit is the release of pollutants, and accidental losses of dangerous chemicals. As we saw earlier in this chapter these 'costs' are now under control and well within budget. PVC is no more environmentally damaging than other plastics. Ideally a PVC plant would take in natural gas, electricity, salt and water, and the only product to emerge would be the polymer and the by-product caustic soda. All waste gases, liquids and solids would be cleaned up and recycled on site. In the next century all chemical plants will be like this, and some have already been built. Admittedly their products would be more expensive but that is a price we should be prepared to pay, and we must keep up the pressure for industry to invest in the necessary technology for site monitoring. The threat to achieving this goal comes from emerging nations, where firms may be able to run a large environmental deficit because of lax regulations, lack of investment, and local people giving priority to jobs over safety and the environment.

Mercury and asbestos are used in the manufacture of chlorine, and these also need to be taken into account on the debit side of the environmental audit. Loss of mercury has been a serious problem in the past, but an environmental target has been set that chlorine plants should lose no more than 2 grams of mercury per tonne of chlorine produced. Some plants can operate with losses as low as 0.25 grams per tonne, and in a typical plant producing 100 000 tonnes of chlorine a year the loss of mercury could be only about a tonne in 40 years. In any case it has been agreed that mercury cells will be phased out by the year 2010. They will be replaced by diaphragm cells which use no mercury. These have asbestos as part of their construction, which is also a debit factor in an environmental audit.

Some chlorine-containing compounds are a threat to the ozone layer. To be so, however, they have to be unreactive towards the lower atmosphere so that they can reach the upper atmosphere and release their chlorine atoms there among the protective ozone which they will help to destroy. The gas vinyl chloride and the volatile liquid ethylene dichloride, both of which are involved in PVC production, are too reactive to survive the journey to the upper atmosphere. They react rapidly with oxygen in the lower atmosphere and are washed out in rain. Minute traces may reach the ozone layer, but these are insignificant compared to the amount of CFCs in the atmosphere. On this environmental score the manufacture of PVC poses no threat to the atmosphere.

Energy and fossil fuels   The chief part of any environmental audit is energy, and most of this comes from fossil fuels. Current methods of power generation based on fossil fuels may be convenient, but many are very inefficient, especially when the waste steam is allowed to escape unused. Other sources such as hydroelectric power are more acceptable, but constructing a dam takes a lot of concrete and energy so even water power is not really 'free'. The chemical industry in general uses only 9% of fossil fuel output, of which plastics account for 5%, and PVC less than 0.5%. The first method of manufacturing vinyl chloride used coal as the hydrocarbon source, in this case to make acetylene gas which was reacted with hydrogen chloride. Some plants in Germany still use this method of making vinyl chloride. Coal reserves are barely touched by PVC manufacture.

Globally we make about 18 million tonnes of PVC a year, which requires 8 million tonnes of fossil hydrocarbons. The industry also uses these as fuel for heat and transport. We need to add together the hydrocarbons used at all stages of PVC production, from the raw materials mentioned above right through to the final product delivered to the consumer. This must include not only the ethylene used to make PVC, but also the energy used to make the ethylene and to generate the electricity to extract chlorine from salt. We also need to take into account other energy that is used in a manufacturing plant for heating and driving equipment, and in the offices and storage depots. PVC has also to be made into various products, all of which takes more heat and power in more factories. And even when the product is finished it still has to be transported to distribution points and thence to shops where the customer can buy it.

In some cases a product made of PVC will last a long time, and longer than alternative materials, so that during its lifetime it could actually save resources. For example, a PVC water pipe will not corrode or burst when frozen, and so will prevent water loss and damage to property. However, it will crack if struck violently, whereas a metal pipe may be able to withstand such a blow. If a PVC component is lighter, and part of a vehicle, it will require less energy to move it. Generally an object made of PVC is about half the weight of one made of metal, so that a car having such a component will use slightly less petrol during its lifetime.

The energy required to produce steel is 32 megajoules per kilogram[13] (MJ/kg), for cardboard it is 56 MJ/kg, and for aluminium is 146 MJ/kg. How do plastics compare with these materials? The energy consumption involved in the production of PVC and other common polymers is as follows: for PVC it is 68 MJ/kg; for polyethylene, polypropylene and polystyrene it is about 100 MJ/kg; for **PET**, the plastic used to make carbonated drinks bottles, it is 123 MJ/kg. The reason why PVC is more economical is mainly because it weighs more, on account of the chlorine it contains.

Having produced the PVC we need more energy to turn it into things, but the energy to do this is generally much less than that to prepare the plastic itself. For example, moulded articles take 10 MJ/kg and sheeting 12 MJ/kg. But energy inputs can be misleading. Sometimes PVC can appear much cheaper, but is not necessarily so,

---

[13] A megajoule is a million joules, and megajoule per kilogram is roughly 100 000 kilocalories per pound.

and in other circumstances it appears dearer even though other cost factors may make it the cheaper alternative in the long run. For example, to make one thousand 250 ml containers from PVC requires 450 MJ; to make them from glass needs 1800 MJ. This shows PVC in a good light, but if the glass container is used several times then this last figure falls considerably. In some cases PVC compares badly at first glance, for example sewer pipes made of clay cost less than half the price of PVC, but PVC pipes are less labour and capital intensive because they can be installed in longer lengths and using lightweight machinery.

Incineration The dangers of fire represent a debit item for plastics when compared to metals and glass. Both natural and synthetic polymers alike are seriously compromised in this respect, and both may contaminate the environment when they burn. Some catch fire dangerously easily, and all produce toxic gases such as carbon monoxide. PVC is among the safest in terms of flammability: paper will ignite at 130 °C, whereas polypropylene requires 260 °C, polyethylene 350 °C, polystyrene 490 °C, PVC 450 °C, and Teflon 580 °C.

When most plastics are ignited they will continue to burn as long as there is oxygen. Most PVC will not burn by itself,[14] unless the air contains 50% oxygen. In ordinary air, which has 21% oxygen, PVC will only burn if it is kept in contact with materials that are already burning. Even then it tends to give out much less heat. Unlike some plastics it does not drip molten material, but it does produce smoke and toxic fumes and it is these which have led to some disquiet. All burning materials produce carbon dioxide and carbon monoxide. Burning PVC produces hydrogen chloride, which is about as toxic as carbon monoxide. On the other hand, burning wool, leather and wood produce the very toxic gases acrolein and hydrogen cyanide.

The risk of producing dioxins is more worrying, but we have seen that PVC does not produce these to any greater extent than say burning wood, or any material which burns to form a hot plasma of carbon, hydrogen, oxygen and chlorine atoms. Such a plasma will

[14] Some flexible PVC with a high plasticiser content will continue to burn once it is ignited.

form dioxins if it cools slowly, and this will be helped by dust particles that are present and which act as a catalytic surface.

Recyling Recycling is also part of the environmental audit. PVC manufacturers recycle their own PVC scrap as a matter of course. Low grade scrap ends up as car mud–flaps or as composition soles for footwear. The building industry uses a lot of PVC in pipes, guttering, flooring, facades, windows, shutters, blinds and duct-ing. All this could be recovered and recycled eventually, and this is beginning to happen. However, PVC waste is still gathered up along with all the other waste and buried in land–fill sites or burnt. That which is buried will slowly degrade over a period of thou-sands of years, but the residues will not leach anything that is dangerous into the ground water. The modifiers it contains will also degrade.

Packaging is often seen by the public as unnecessary waste, and many concerned people are willing to take the time and trouble to sort such waste for recycling. A mixture of plastics can be shredded and mechanically sorted according to the different densities of the various plastics. PVC and PET, which is also used to make bottles, are almost identical in this respect, but if a mixture of the two is passed through a hammer mill then the PVC is fragmented more than the PET, and the mixture can be separated by sieving. The PVC stream will contain small amounts of PET (about 0.5%), but this acts as an inert filler.[15]

To help sort waste plastics they are stamped with identification codes, but separating them by hand is slow, costly and prone to error. Automatic sorting can be faster, cheaper and foolproof. A method has been developed at the Sandia National Laboratories in Albuquerque, New Mexico, which relies on an infrared ray being reflected from a piece of plastic and then analysed by computer. This can distinguish the six most commonly used plastics: PET, PVC, polystyrene, polypropylene and the two types of polyethy-lene, the low density form used in making plastic bags and the high density form used to make things like containers.

---

[15] However the 0.5% PVC in the PET stream renders this plastic useless.

To summarise: our environmental audit would suggest that PVC is no worse than other plastics or natural materials, and in some ways it scores well: it is cheap; not very flammable; can be recycled; and many of its uses help save lives. In other ways it loses out: it depletes fossil reserves; and some dangerous materials, such as dioxins, are formed during its manufacture and in its incineration—but as we shall see in the next chapter, the amounts are insignificant compared to other sources of dioxin. On balance, if PVC were to be phased out and replaced by another material that was of biological origin, this would certainly be to the detriment of human society, but a slight benefit for the environment. I still think the former far outweighs the latter.

PVC is the most versatile plastic there is. It can be turned into anything from a park bench to a length of tubing so fine and gentle that it can be threaded into an artery of a sick baby. PVC can be as transparent as crystal, or as black as coal. It can be any colour and any shape, it does not rot, splinter or shatter, and it can automatically be sorted out of waste plastic and recycled. In the next century we could see PVC descending through a cascade of uses, starting perhaps with clear plastic bottles, which could be recycled as coloured containers, which in their turn could be turned into garden fencing, office floor-ing or sewer pipes. Finally, when its quality was too poor for recy-cling, it could be incinerated to provide heat energy and its ashes buried. We are still a long way from such careful stewardship of the Earth's resources, but we can help our children to plan for such a day.

At the start of this chapter I said that PVC was accused of causing cancer, contaminating our food and polluting the environment with dioxins. You will now know why I said that all these were true, and I hope you will now understand why I also said we need not worry about them. As far as I am aware, no member of the public has ever been harmed in any of these ways by PVC, and many people owe their lives to it. It is time we learned to live in peace with a rather wonderful plastic.

CHAPTER SEVEN

Dioxins, The World's Deadliest Toxins?

Chlorinated phenols; tracking down dioxins; dioxins at large; hidden dangers; the verdict

IN the previous chapter I said that one of the fears about PVC concerned the emission of **dioxins**,[1] because these could be formed as this plastic was burnt in incinerators. We now realise that these fears about PVC were exaggerated, but the threat from dioxins is still taken very seriously, whatever their origin. Every time dioxins are mentioned in the media they are described by environmentalists as the deadliest chemicals ever made, and to prove the point people are told that less than a millionth of a gram is enough to kill a guinea pig. They may also be told that dioxins cause cancer and birth defects. It is therefore not surprising that people worry when dioxins are shown to be present in parks and swimming pools, or when they turn up in paper and food; and when dioxins are found in mother's milk, and in our own bodies, we begin to protest and campaign that whatever is producing them must be stopped.

As a consequence of some alarms about dioxins, people have fled their homes in panic, such is the fear that they provoke. And yet no member of the public has ever died of dioxin poisoning, despite the fact that for 40 years the chemicals industry inadvertently produced large amounts of dioxins as impurities in other products. During this

---

[1] A reminder that when the name of a substance appears in **bold** type, it means that there is more about its chemistry in the Appendix.

time there were some major industrial accidents and explosions that released dioxins, yet the number of workers who died from exposure to them was probably only four. How can a chemical be so deadly and yet this death toll be so low?

The amount of dioxins in the environment increased markedly when the chemicals industry started to make organochlorine compounds on a large scale early this century, and especially when chlorinated weedkillers came into general use. It was always known that dioxins were trace impurities in such weedkillers, but they were not regarded as particularly dangerous, and this seemed to be borne out by the apparently safe application of these garden chemicals for over 30 years. The occasional worker in industry and agriculture was sometimes adversely affected, but that is only to be expected considering the risks involved in dealing with large amounts. The risk to the public was regarded as negligible, and seemed worth taking as part of the price of more efficient farming in the years after the Second World War.

Dioxins are organochlorine compounds, and today they stand condemned by some environmental groups on this fact alone. Greenpeace, for example, has declared war on all organochlorines, and they refer to chlorine as 'the Devil's element', with some justification; some of their major struggles have been fought over organochlorine compounds such as the insecticide **DDT**, and the aerosol gases, the **CFC**s. To a chemist like me it seems a little unscientific to vilify a common chemical element, or to rail against a chemical bond, in this case the carbon-to-chlorine bond, the possession of which determines whether a material is an organochlorine compound.

Living things make organochlorines. The red algae that live in the sea produce chloroform and carbon tetrachloride in vast amounts, seemingly unaware that these are environmental pollutants and health hazards, and that they are now banned for use as anaesthetics or solvents. Hundreds of other plants and microbes also make organochlorine molecules, and these have also been detected in the gas clouds of volcanic eruptions. Nor are all synthetic organochlorines bad. Some bring enormous benefits, such as the versatile plastic PVC (as we saw in Chapter 6). The effective healing drug aureomycin is also an organochlorine chemical.

Ironically it was chemists themselves who put dioxins at the top of the agenda of public debate, not because they produced them, but because they found ways of measuring them in incredibly tiny amounts. Advances in chemical analysis made it possible to detect one dioxin molecule in a million billion molecules of water, which is like finding a needle in a haystack the size of a mountain. While this was a triumph of the chemists' skill, it opened a Pandora's box, because it suddenly made people aware that dioxins were everywhere. They were seeping into water courses from landfill sites where the chemicals industry had dumped its unwanted wastes; they were polluting the countryside around Seveso in Italy where they had been scattered in an explosion at a chemical works in 1976; they were in the soil of Vietnam where millions of litres of dioxin-contaminated defoliant had been sprayed. Then people discovered that dioxins were in their food, in the air they breathe, in the water they bathe in, in the paper tissue they used to wipe their nose, and in the disposable nappies (diapers) they were putting on their babies. Even mother's milk was discovered to harbour them. They were there in our own bodies. How had it all happened? Where was it all coming from? What had the chemicals industry been doing?

In fact dioxins have been part of the environment since dinosaurs walked the Earth. There is a natural background level of these chemicals which are formed whenever things burn. The first bush fire or forest fire polluted the planet with them and these natural sources have been producing them ever since. What the chemicals industry had done was add a lot more, and they may even have doubled the amount in the environment over a relatively short period of time. Although this source of the pollution is under control we still cannot relax our guard, because other things are still producing dioxins. However, this was not realised for a long time.

Chlorinated phenols

Let me begin the dioxin story by going back to where it apparently started—with a chemical called **phenol**. This used to be known as carbolic acid, and comes as crystals that are slightly soluble in water. Phenol came to public notice as the first surgical antiseptic, pioneered by Joseph Lister at Glasgow University, Scotland, in 1865. He showed that if phenol was used in operating theatres to sterilise equipment and dressings there was less infection of wounds, and patients stood a much better chance of survival. By the time of his death, 47 years later, Lister's method of antiseptic surgery was accepted throughout the world.

As a result of Lister's success, phenol became a popular household antiseptic and was put as an additive in so-called carbolic soap. Despite its benefits at the time, this soap is now banned, and you will discover why if you look up phenol up in the latest edition of Sax's *Dangerous Properties of Industrial Materials*, where you will find frightening phrases like 'spillage on the skin can result in death within 30 minutes', 'kidney damage', 'toxic fumes' and 'co-carcinogen'. Clearly phenol is totally unsuitable for general use, but the benefits 130 years ago plainly outweighed the disadvantages. However, it was not long before phenol was turned into something much safer: chlorinated phenol. Chlorine gas reacts with phenol to add one, two, or three chlorine atoms to the molecule. The two-chlorine derivative is called **2,4-dichlorophenol**, and the three-chlorine derivative is called **2,4,6-trichlorophenol**. These chemicals are less soluble in water than phenol itself, but what does dissolve is a much more potent antiseptic, so far less is needed to do the job.

For the first half of this century a bottle of antiseptic chlorophenols was to be found in the medicine cabinet of most homes, and the solution was used for bathing cuts, cleaning grazes, rinsing the mouth, and gargling to cure sore throats. Little did our grandparents realise that they were doing all this with a solution that probably contained dioxins. The manufacture of chlorinated phenols became an even more flourishing part of the chemicals industry when other remarkable properties of these compounds were discovered.

Chemists can ring all sorts of changes with phenol and chlorine. There are in fact 19 different chlorophenols, and other groups of atoms can also be added to these molecules. One product in particular, with the name **2,4-dichlorophenoxyacetic acid**, or **2,4-D** for short, had a curious effect on plants: it acted like a growth hormone, but one that was out of control. Plants that absorbed 2,4-D died, and it made a particularly effective weedkiller against broad-leaf weeds. A tiny drop of this herbicide was all that was needed. Yet it did not kill grass or crereals, so that it became a superb selective weedkiller for lawns and grain crops. In some circumstances 2,4-D was used to trick plants into flowering, which was how visitors to Hawaii could be guaranteed a greeting of pineapple flowers all the year round.

But was 2,4-D safe? Tests on animals showed that it was, and for rats the toxicity measured as the LD_{50} factor[2] was about 0.5 gram per kilogram of body weight. Scaled up to human terms this would mean that 35 grams would be a likely fatal dose for the average person who weighs 70 kg (155 pounds). Because it required only 100 grams of 2,4-D to treat a hectare of land (about two and a half acres), and because the solution in which it was applied was very dilute (about 100 parts per million), it would have required the drinking of 350 litres (90 gallons) to take in a fatal dose. Moreover 2,4-D was ideal for weeding public places such as play areas, paths and parks because it could be sprayed from a back-pack. It was cheap, effective, more selective than other weedkillers, and much safer than the sodium arsenate and sodium chlorate that had been popular weedkillers up to the 1950s. Sodium arsenate had featured in many domestic murders because it is a deadly poison, and sodium chlorate had featured in accidental fires and explosions, because it is a powerful oxidant.

But 2,4-D did not kill every broad-leaf plant, and not the woody creepers and brambles that choke woodland paths and encroach on railway lines. The answer to those was to add another chlorine to the molecule to produce **2,4,5-trichlorophenoxyacetic acid (2,4,5-T)**, and this herbicide came on to the market in 1948. It appeared to be safer than 2,4-D, even though we now know that it must have contained larger quantities of dioxin. Rats did not die when they were fed a diet rich in 2,4,5-T for months, at what would have been the equivalent of a human eating 2 grams a day, but they suffered severe damage to their liver and kidneys as we might expect.

2,4,5-T contained more dioxin than 2,4-D because it was made in a different way. With 2,4-D you start with phenol and add chlorines, but with 2,4,5-T you start with chlorinated benzene and convert it to a phenol. A subsequent chemical reaction converts it to 2,4,5-T, and this step is carried out at 140 °C. The process has to be very carefully watched because if the temperature

---

[2] The LD_{50} is explained in the Box on page 141.

rises above 140 °C a side-reaction takes place and this produces dioxin.

The makers of 2,4,5-T were always aware of this possibility and took great care to see it did not happen. They were not aware then, as we are today, of the dangers of dioxin, but they knew that any product contaminated with it was unpleasant to handle because it caused a skin complaint called chloracne. The victim's face became covered with blackheads and cysts, particularly around the eyes and ears.[3] Chloracne generally clears up within a few months, but some people continue to be affected by it for many years. The German dermatologist K.H. Schulze first realised the link between chloracne and dioxin when he had a patient with the disease who was a chemist, and who had made a 20 gram sample of pure dioxin in the laboratory.

Apart from the chloracne, which developed when workers were exposed to large amounts, dioxin did not appear to be toxic to animals in the tiny amounts in which it was present in 2,4,5-T. There were about 10 parts per million (ppm) of dioxin in the product and this was deemed to be acceptable. Had they been manufacturing 2,4,5-T under today's more stringent laws, then the amount of dioxins permitted would have been set at least ten thousand times less than this at 1 part per billion (ppb), although ideally the herbicide should contain none at all.

Back in the 1950s the benefits of 2,4,5-T were all that mattered. It killed the tough weeds that 2,4-D did not kill. Indeed, it was so successful in killing woody plants that it was deployed in the Vietnam War. For about seven years, from 1962 to 1969, a 50:50 mixture of 2,4-D and 2,4,5-T called Agent Orange was sprayed from the air to destroy the dense foliage of trees that provided cover for the Viet Cong troops. During those years 50 000 tonnes of the herbicide were sprayed on the jungle. Agent Orange was contaminated with 2 parts per million of dioxins, and as it rained down on the forests it carried with it a total of 100 kilograms in all. The first inkling that something other than defoliation was happening came with reports that deformed babies were being born in higher numbers than

---

[3] Other chlorinated compounds, such as the poly(chlorinated biphenyls) (PCBs), can also cause chloracne.

expected in zones that were heavily sprayed. Nor was this just Viet Cong propaganda. We now know now that dioxins are teratogens, compounds known to cause birth defects.

The US Government put restrictions on the domestic and agricultural use of 2,4,5-T in 1970. The public were still not alarmed, but those in the industry knew by then that things could go seriously wrong with the manufacture of 2,4,5-T, although they were still unaware when this happened that the dioxins which were produced were so dangerous. However, before we look at the social and environmental issues surrounding dioxin we need to look more closely at the chemical itself.

Tracking down dioxins

From a chemist's point of view it seems rather ironic that the dioxin scares were based on improvements in chemical analysis. As analysis got better, the fear got worse. The alarms were magnified by the confusion over the term dioxin. Let me explain: the word 'dioxin' is used not only for one particular chemical, but also for a whole group of very similar chemicals. Over the years this has led to confusion. We can see the problem if we consider the following statements, both of which are true: (i) dioxins are among the deadliest molecules ever made; and (ii) dioxins can be completely non-toxic. In fact dioxins are a group of chemicals each having two benzene rings joined together through one or two oxygen atoms. Those with two linking oxygens are called dibenzo-dioxins (DD for short), and those linked through one oxygen are called dibenzo-furans (DF for short). Often they occur together, and the general term dioxins is used for the whole set.

Both DD and DF can have up to eight chlorine atoms attached to the molecule. There are 75 ways of arranging between one and eight chlorines on the dibenzo-dioxin molecule, so there are 75 separate compounds in this group of DD dioxins. There are 135 ways of arranging between one and eight chlorines on the dibenzo-furan molecule, so again there are 135 separable compounds in this group of DF dioxins. Of the 75 dibenzo-dioxins only seven are toxic, and of the 135 dibenzo-furans only ten are toxic. Out of a total of 210 dioxin-type molecules only 17 or 8% are poisonous; the other 92% are relatively harmless.

The most dangerous of all are 2,3,7,8-tetrachlorodibenzodioxin, abbreviated to 2,3,7,8-TCDD, and often referred to just as **TCDD** and 2,3,7,8-tetrachlorodibenzofuran or **TCDF**. The second most toxic are 1,2,3,7,8-PCDD and 2,3,4,7,8-PCDF, both of which have half the potency of TCDD.[4] The other 14 toxic derivatives have at most only about a tenth of the potency, many have only a thousandth, and the rest we can regard as essentially non-toxic. The total toxicity of a sample of dioxins is given by adding together all the toxic components, taking into consideration their relative amounts and toxicities. TCDD and TCDF get a toxic rating of 1, others 0.5, yet others 0.01, and some even 0.001. These are combined to get the so-called toxic equivalent, or TEQ.

TCDD is *the* dioxin. There is no doubt that this is a deadly poison, and just how deadly can be seen from the table opposite, which shows the LD_{50} for several animals.

To put these toxic doses into perspective, and to compare them to other poisonous chemicals, we should note that for humans the LD_{50} for botulinus toxin is 0.01 milligrams per kilogram, while for nicotine it is about 1000 mg per kg. As the table opposite shows, the fatal dose of TCDD for hamsters is about 5 mg per kg bodyweight, but a guinea pig needs only one *micro*gram, which is a millionth of a gram. It was this figure which led to TCDD being commonly referred to as 'the most toxic synthetic chemical known', which is probably correct—at least for guinea pigs—but as we saw on pp 140–1 these poor animals are rather sensitive to poisons. Although the LD_{50} figure for humans will probably never be known, we might at first glance expect it to be like those for our primate cousins, the monkeys. However, experience suggests that it is much higher than this and maybe even much higher than the LD_{50} of the hamster.

TCDD has several effects on animals apart from killing them. It will damage the thymus, spleen and testicles, enlarge the liver, build up fatty deposits, cause cancer and lead to birth defects. Research at the University of Wisconsin has shown that TCDD does not affect

[4] The initial letter T in these abbreviations stands for tetra, meaning four, and P stands for penta, meaning five.

The toxicity of TCDD

Animal	LD_{50}*
Guinea pig	0.001
Monkey	0.070
Rat	0.200
Dog	3.0
Hamster	5.0

*milligrams per kilogram body weight.

sexually mature male laboratory animals, but if it is given to pregnant rats then their male offspring showed reduced sexual drive and had smaller sexual organs. Other research shows that TCDD affects the immune system. The toxic response of an animal to TCDD can vary widely, as the LD_{50} shows, and this is true of other symptoms—for example humans break out in spots when exposed to it, but rodents never do.

What alarms people most about TCDD is the threat of cancer. When the US Environmental Protection Agency ordered tests on laboratory animals it discovered that TCDD was the most potent cancer-causing chemical it had ever encountered. Was the same true with regard to its causing cancer in humans? Happily the answer appears to be no, judging from the fate of those who have been accidentally exposed to large amounts of it in Europe and the USA. The numbers involved have thankfully been relatively small. The circumstances under which they were exposed to TCDD were very varied, but this has meant that there is very little evidence linking the extent of exposure to the incidence of disease. Nevertheless, some conclusions can be drawn, the main one being that TCDD appears to increase the risk of some rare forms of cancer, but only very slightly. It may even *protect* women against breast cancer.

Surveys have been carried out on workers in chemical plants which made products that were contaminated with TCDD. Some people had been exposed to dioxins for over 20 years of continuous employment at companies where 2,4,5-trichlorophenol was produced. The number who had eventually died was more or less the number expected from natural causes, although the incidence of cancer was indeed slightly higher than expected when compared with the general population, but only by one or two, and with no particular type of cancer being prevalent. One study was based on 1500 workers who had been employed at a chemical plant in Hamburg where there had once been an accident in the 1950s. Of those who had worked there for at least three months, a few more than statistically expected had died of cancer, but again we are talking of single figures.

An American study carried out by Marilyn Fingerhut and colleagues at the National Institute for Occupational Safety and Health in Cincinnati examined the health records of over 5000 male workers at dozens of chemical companies where products were produced that were contaminated with TCDD. Men who were exposed to very little dioxin did not have significantly increased levels of any kind of cancer, but among those who were heavily exposed there were a few extra cases of soft tissue sarcoma.

Another large study involved over 18 000 workers from ten countries who had either worked on the production of trichlorophenol or had sprayed 2,4,5-T as a herbicide. There was a doubling of deaths from cancer of the soft tissue, which sounds rather worrying until you realise that the investigators found that four people had died of this condition over a period of 20 years, instead of the expected two. We will look more closely at the long term analysis of those affected in some dioxin disasters later in this chapter.

TCDD may be a carcinogen for laboratory animals but not for humans to any extent, and clearly our defences can generally repair the DNA damage it causes. If TCDD has a role to play it is not so much as an initiator of cancer, but as a factor in its later development. Even then it is not easy to understand what this might be. TCDD is also an anti-cancer agent: animal studies revealed a slight increase in liver cancer, but a remarkable *decrease* in breast and uterus cancers to animals fed this dioxin.

Chemical analysis

The aspect of dioxins that has clouded the issue, and indeed led many campaigners astray, has been their remarkable sensitivity to chemical analysis. This lay behind many of the false alarms. There could have been no environmental dioxin scare in the 1950s because chemical analysis at the time could only have detected them at levels of parts per thousand. By the mid-1960s accuracy had improved a thousandfold and chemists could detect parts per million. By the mid-1970s another thousandfold increase made it parts per billion, and by the mid-1980s yet another thousandfold jump made it an unbelievable parts per trillion.[6] Today it is even possible to identify the components in a mixture of dioxins. Previously chemists thought there was only one dioxin being produced in the manufacture of 2,4,5-T, and that was the deadly TCDD. Now we know that there are others as well. The peak that recorded 'dioxin' on an analyser chart 20 years ago might well have been a composite of 20 dioxins or more, many of which were not toxic at all, and yet which were being recorded as though they were TCDD and assumed to be equally dangerous.

Professor Christoffer Rappe of the University of Umeå, Sweden, has made a close study of how to analyse dioxins and is the world authority on the subject. Several methods are available, but the best is the so-called capillary gas chromatography method which separates the component dioxins, and high resolution mass spectroscopy which can identify them individually. Extreme skill is required for measuring amounts as small as a billion millionth of a gram, and even today there are only a few laboratories that can do it reliably.

Analytical results for dioxins are displayed as a chart on which there are clusters of many signals as we might expect, and in theory all 210 dioxins might be present. It is possible to identify the dioxin of each signal by injecting a sample of that particular dioxin, such as 2,3,7,8-TCDD, which has been tagged in some way, say by having its atoms as the carbon isotope carbon-13 instead of the normal carbon-12 isotope. All the toxic dioxins are now available as labelled versions using atoms of carbon-13, carbon-14 (radioactive), chlorine-37 and hydrogen-3 (which is also radioactive). By having a heavier version of an atom in a molecule we can identify it by mass spectrometry which measures its molecular weight. These days we no longer expect to hear of analysts merely finding 'dioxins' in something, and so worrying us needlessly. Now we expect them to say how much of these

---

[6] In weight terms these quantities are as follows: one part per thousand is one gram in a kilogram. One part per million is a gram in a tonne. One part per billion is a gram in a thousand tonnes, and one part per trillion is a gram in a hundred thousand tonnes. Even this last amount does not represent the limit of modern chemical analysis, which can detect dioxins in so-called femtogram quantities, which is equivalent to a gram in a million tonnes.

dioxins are the poisonous TCDD and TCDF, and what is the overall TEQ (p180).

Analytical chemists were the ones who finally set the dioxins in their global and historical context. In 1991 they found dioxins in soil collected 150 years ago, long before the organochlorines from the chemicals industry had been let loose on the world. There in the carefully preserved samples of yesteryear were the dreaded TCDD and TCDF, along with many of the rest. The total amount of dioxins in the 1840s soil samples was tiny, on average about 30 picograms per gram of soil (30 parts per trillion), but they were undeniably there. The soil samples were taken from Rothamsted, a rural area in the south of England, but as this country was already heavily industrialised in the 1840s the dioxins could still have come from human activity, such as the burning of coal, although we know now that this type of combustion produces relatively little.

Other samples that predate the industrial age also contain dioxins, and they have been found in sediments in Japan that were laid down over 8000 years ago. Clearly dioxins have been released into the environment throughout history. When any wood burns dioxins are formed, and when forests burn a lot of dioxins are formed— enough to contaminate the whole planet. Wood is still gathered and burned as fuel by people in the Third World, and forests fires are started by lightning strikes or other accidents. How much wood gets burnt is not certain, but it could be as high as 5 billion tonnes per year, and if we assume that one ton of wood produces a milligram of dioxins (this is the amount generally produced by incinerating organic waste) then this source may release 55 *tonnes* of dioxins into the environment every year. The planetary burden of dioxins has risen in this century, partly because of the chemicals industry, and partly because of the increased combustion of all kinds of fuel. The level in Rothamsted soil in the early 1980s was around 90 picograms per gram—three times the level of the previous century.

These days samples for analysis are regularly taken from inside chemical plants and tested for dioxins. Monitoring is done not only on batches of chemicals, and on the products made from them, but also on waste effluents and emissions, and on soils and sediments both inside the plants and in the surrounding areas. Workers and locals are checked, and so is wildlife. Some chemical works that use chlorine gas in their processes now do environmental audits, and are able to say exactly how much dioxin they produce each year and how much escapes. The amounts are tiny. Whereas a manufacturer of 2,4,5-T in the 1960s might have sent out 1000 grams of TCDD each year mixed with their products, today a chemical firm would be worried if it lost only 10 grams a year.

Analysis for traces of dioxins is one thing, but experimenting with these chemical compounds in the laboratory is another. Because TCDD is so toxic there have been few chemists willing to research the chemistry of this key dioxin, but a few brave souls have investigated it. From their researches we know that it is a very stable molecule, but it will react with some chemicals. Ultraviolet light will knock off its chlorines, and if it is heated above 750 °C it rapidly decomposes.

The natural process that destroys most dioxins in the environment is probably the action of sunlight, although there are some microbes that can digest them. Dioxins are slightly volatile, and when they enter the atmosphere they are then at the mercy of ultraviolet rays from the Sun. Dioxins are resistant to breakdown in animals, and tend to collect in fatty tissue. There is some evidence from rats and dogs that they are capable of converting them to simpler molecules that are then easily lost. There is also evidence that enzymes that are present in sewage can remove dioxins.

Dioxins at large

The first industrial accident involving dioxins occurred in 1949 in Nitro, West Virginia, at a Monsanto plant which was making 2,4,5-T. Over 120 workers were severely affected by it and developed chloracne. Those who were exposed to dioxin were monitored for 30 years by the University of Cincinnati's Institute of Environmental Health, and were found to live longer than average, and to have fewer cancers and other chronic diseases than normal.

The second accident happened on 17 November 1953 at the BASF plant in Ludwigshafen, Germany, where trichlorophenol was being made. The temperature of the reaction went out of control and an explosion occurred which released dioxins over the site and the immediate neighbourhood. Of the 250 workers who were exposed to dioxin, about half suffered chloracne. Attempts were made to clean up the site by washing with detergents, and when this had little effect, they resorted to burning contaminated material and blow-torching surfaces. When this failed to remove it all they demolished the entire building. The medical histories of the workers who were affected by the accident were investigated 25 years later, by which time eight had died of cancer, five more than statistically expected. By 1990 the number who had died was 78, of whom 23 had cancers of various kinds. Again this was higher than would normally be expected for a group of this size among the general population.

The third accident was in 1963 at the Philips plant at Duphar in the Netherlands. Of the 50 people put to work to clean up the site, 16 went down with chloracne, and not long afterwards four of them died. These are the only people who may have died as a result of direct contact with dioxins. The plant was sealed for ten years before being dismantled brick by brick. The rubble was embedded in concrete blocks and sunk in the Atlantic Ocean.

The fourth industrial accident occurred in 1968 in the research laboratory of Coalite Chemicals in the UK. The chemical reaction making 2,4,5-T went out of control and the vessel holding it exploded, killing one man. The building was not seriously damaged and continued in use until people began to go down with chloracne. Eventually about 80 people were affected. The company then dismantled the whole plant and buried it down an old mine shaft. A new laboratory was constructed, equipped with built-in safety features, and when another accident occurred in 1974 these played a major part in containing it.

Then in 1976 came the disaster that brought dioxins to the attention of the world.

Seveso, Italy, 1976

At 7.30am on Saturday 10 July 1976, a cloud of gas and dust escaped when a reaction vessel overheated in the Icmesa chemical plant, about 13 miles north of Milan. The accident was caused by a sequence of failures, both mechanical and human. The pressure inside a reaction vessel making chlorophenols shot up, and a safety valve came into

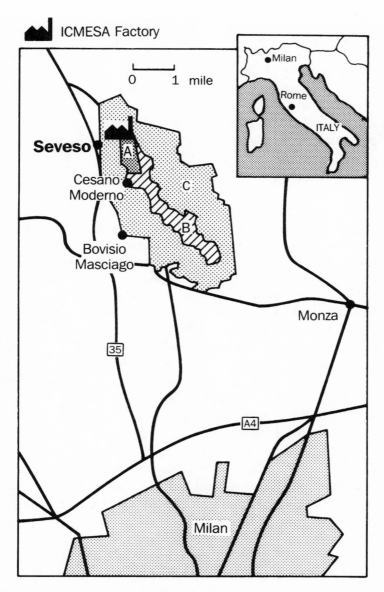

The Seveso accident. The diagram shows the fall-out zones of dioxin contamination following the release of material from the Icmesa factory on 10 July 1976. The degree of exposure was calculated according to the amount of dioxin in the soil. Zone A, nearest the factory, had the highest levels, and 724 people lived in this area. Zone B, with lesser amounts, extended in a narrow band for about 4 km (2.5 miles), and 4800 people lived in this area. In Zone C, which showed a small amount of contamination, there were about 31 000 residents. [Source: *Journal of the American Medical Assocation.*]

operation. The solvent in which the reaction was occurring was **ethylene glycol** which boils at 197 °C, whereas normally the reaction is done in methanol which boils at a lower, safer 65 °C. The disaster might still have been averted if the safety valve had released escaping materials to a catch-pot, which is what usually happens in such situations. Instead a cloud of gas and fine dust shot into the atmosphere, and this settled on the nearby suburb of Seveso with its population of 17 000. Among the dust was an estimated 3 kilograms of TCDD.

The need to evacuate the area was paramount, but the carrying out of that evacuation was less than efficient. Attempts were made by those at the chemical plant to alert the community, but key local officials were away on holiday for the weekend. When they were contacted they did not appreciate the gravity of the situation, and for about a week no action was taken. Meanwhile, pets in the town sickened and died, and the hands and faces of the children who played with the 'snow' began to break out in chloracne.

Squabbles between the company and local officials wasted valuable time, but eventually 250 people were evacuated. Of these 180 had by now contracted severe chloracne. There were 150 pregnant women who had been exposed to dioxin, and 30 of them decided to have abortions for fear that the children they were carrying would be born badly defective. It turned out that 29 of the foetuses were perfectly normal, and one had Down's syndrome, which is not caused by TCDD but is a defect that occurs at the moment of conception. Of the 100 women who came to term, only two babies had birth defects, which is well within the number normally expected, and both were corrected by surgery.

Eventually 600 people were told to leave the Seveso area that was most affected, and 2000 more were given blood tests. Despite the heavy exposure to dioxins, no-one died, demonstrating that compared to animals, humans are not particularly sensitive to TCDD. The Seveso affair rumbled on, sometimes with bizarre results. For example, contaminated waste from the area 'disappeared' only to turn up years later in a disused abattoir in the village of Anguilcourt-le-Sart in France. In 1983 five senior managers of Icmesa were brought to trial and given prison sentences of up to five years, while the parent company, Hoffmann-La Roche, paid out £80 million in compensation. Analysis at the plant revealed that the reactor which had gone out of control had actually been shut down for the week-end, but it was later realised that the process itself was capable of heating up spontaneously.

In 1988 a more detailed report on Seveso came out, and this focused on the 15 291 births that had occurred in the area since the accident in 1976. Among the Seveso babies born after 1976 there were 742 with defects, which is about 5%, and the number to be expected statistically for such a population as a whole.[7] The 26 women living nearest the plant at the time of the disaster and who were most exposed to TCDD gave birth to perfectly normal babies. Food grown near the Icmesa factory was analysed a year after the accident and no TCDD was detected in the flesh of apples, pears or peaches, but it was detected at 100 parts per trillion in their peel, suggesting that the source was dust and not uptake by plant roots.

The University of Milan's Institute of Occupational Health (IOH) has monitored the Seveso disaster. Medical records covering the ten years following the accident were analysed carefully, but there were no dramatic rises in cancer cases, even in the most heavily contaminated areas of the town, just one or two more cases than might have been expected for a population of around 35 000. Among women there were more cases of cancer of the gall bladder, but fewer cancers of the stomach, colon, breast and uterus. Men had a higher incidence of hematologic and liver cancers, but fewer cases of stomach, colon and larynx cancers.

In 1993 Pier Alberto Bertazzi of the IOH published a more detailed analysis of the hospital records covering the years following the disaster, and compared the incidence of various cancers in 37 000 people who were exposed to dioxin and 182 000 who were not. In the exposed group there were slightly more cases of some rare types of cancer, but the numbers were in single figures. There were no extra cases of cancer among the 724 people who suffered the greatest degree of exposure to dioxins. The extra cases of cancer came from the 4800 people who were in the 'moderately contaminated' area and the 32 000 people in the 'least-contaminated' area.

---

[7] The majority of these were simple birthmarks.

Love Canal, New York, 1978

Love Canal is near Niagara Falls in New York State. It was a community of about a thousand homes, some of which were built on top of an old chemical waste tip, seepage from which was leaking dioxins and other materials into the canal. Luckily the solubility of dioxins in water is very low, and a cubic metre of water (1000 litres) will dissolve only 0.2 milligrams (giving a solution concentration of 0.2 ppb). Exactly which dioxin derivatives were in the canal could not be determined at the time; it was known only that they were 'dioxins'.

Those living nearest to the canal itself were convinced that their health was being affected in many ways. Some even blamed dioxins for causing their asthma and piles, while others thought that chemicals were more of a threat to the next generation, as evidenced by miscarriages, deformed babies, and puny children. There were fears that the pollution was causing cancer and brain damage, the latter being evidenced by cases of epilepsy and unexpected suicides.

Action was needed. The New York State Commissioner of Health showed the way, and published a report with the alarming title *Love Canal: Public Health Time-bomb*. It exploded immediately. Over 230 families living nearest to the canal demanded to be rehoused. Studies were initiated, and those residents who said they were affected filled in a questionnaire. The Environmental Protection Agency (EPA) got involved, and worried about possible genetic effects and damage to the central nervous system caused by the chemicals. When in 1980 they reported that there was a real danger, President Carter declared Love Canal an emergency area and the remaining 700 or so families were evacuated. The canal was dredged and the contaminated sediment sealed in drums and taken away. It was a triumph for community action groups, and in 1990 Lois Marie Gibbs, a local housewife and environmental activist, was awarded $60 000 for her work by the Goldman Environmental Foundation of San Francisco.

Further research continued, but somewhat perversely the new scientific and medical evidence did not support the claims that the leak of dioxins was causing health problems. The Center for Disease Control (CDC) had been naturally alarmed by all that had happened and carried out its own survey, comparing Love Canal residents with a similar group chosen from a community far removed from the area.

The CDC analysis was conducted on a so-called double-blind basis, which eliminated subconscious factors that might influence the results. The survey found that the illnesses afflicting the residents of Love Canal were not unusual, but were to be expected in a normal community of that size.[8]

Times Beach, Missouri, 1983

In 1971 a chemical plant in Verona, Missouri, that produced 2,4,5-T hired a company to clean out sludge and other oily waste from the plant. The company found a second use for all this, and spread it on horse arenas throughout the state, where it compacted the dust and loose earth into a hard surface. This treatment also had an added bonus: as well as keeping down the dust, it also kept down the flies. The sludge was contaminated with dioxins to which horses are particularly sensitive, and they began to sicken. Over 50 died. In 1974 the CDC investigated and found the cause to be the arena soils which contained 33 ppm of dioxins.

That appeared to be the end of the mystery, but the waste had also been sprayed on the dusty roads of Times Beach, and that was another time bomb that would explode eight years later. When the nearby Meramec River flooded in December 1982, it washed dioxin-contaminated dirt into homes and shops. A team of EPA officials, dressed in protective clothing, arrived to take samples and assess the danger. The locals fled. In the end the US Government bought the whole town for $33 million and evacuated it. Most of the alarmed residents were happy to quit and to be tested for dioxins. They were reassured to find the levels in their blood were no higher than in the rest of the population, and probably never had been. A closer investigation of the incidence of cancers and birth defects in the people of Times Beach found them to be no more common than elsewhere.

---

[8] According to the *New York Times* of 26 July 1990 the homes at Love Canal have been refurbished and new residents have moved in. In 1994 a Federal court in New York State ruled that the chemical company which dumped the waste chemicals at Love Canal in the 1940s and 50s was not liable for punitive damages.

Agent Orange

For over seven years the US Air Force sprayed Agent Orange on the jungles of Vietnam. The 50 000 tonnes (about 40 million litres) they used during that time contained about a tenth of a tonne of dioxin as contaminant. Could this have affected the men who carried out the operation, and those on the ground? Veterans of the campaign who had been involved with Agent Orange were tested many years later and found to have only 5 parts per trillion of dioxins in their blood, but those who loaded the herbicide on to the aircraft had higher levels. Ex-soldiers had on average only slightly more dioxin in their bodies than the general public, for whom the normal range is 2–20 parts per trillion.

Nevertheless, veterans suspected that dioxins might be causing various illnesses and they began to agitate for compensation, although some of the effects they attributed to dioxins, such as baldness and impotence, had never before been linked to these chemicals. A major law suit was started in 1984 and this cost the chemical companies who had manufactured Agent Orange $180 million in an out-of-court settlement. Closer analysis failed to reveal an unusual pattern of disease, and though the CDC tried very hard to find a correlation between Vietnam service and health problems, they were not able to pin any condition down to dioxin exposure. The difficulty was in knowing which of the three million US troops who served in Vietnam had been exposed to Agent Orange, and the extent of individual exposure.

Part of the problem was again due to chemical analysis, because for many years it was very difficult to measure the amount of TCDD in someone's body. Up until 1986 a person had to donate 20 g of fatty tissue for analysis, and this had to be removed in hospital. Then a method of measuring the level of TCDD in blood serum was developed, so that a patient needed to give only a pint of blood, whereupon veterans came forward in large numbers for testing. About a quarter of a million men were examined and 27 000 were admitted to hospitals for illnesses thought to be related to dioxins. The only reliable sign of exposure to dioxins like TCDD is chloracne. Only five Vietnam veterans had fallen victim to this disease and qualified for disability payments. The men whose job it had been to load aircraft with Agent Orange and to clean up afterwards were deemed to have been most exposed, and they have been selected for monitoring until the year 2002. So far this study has shown that the half-life of TCDD in the human body is about seven years, and much longer than presumed.

In 1993 the report entitled *Veterans & Agent Orange: health effects of herbicides used in Vietnam* was published. This was a review of the 6400 scientific papers published on Agent Orange and had been carried out by the Institute of Medicine (IOM), part of the US National Academy of Sciences complex. It said that five medical conditions show a statistical association with the herbicide: soft tissue sarcoma; non-Hodgkin's lymphoma; Hodgkin's disease; chloracne; and porphyria cutanea tarda, which is another type of skin disorder in which blisters form. Whether these were caused by traces of TCDD, or by the herbicides themselves in Agent Orange (2,4-D or 2,4,5-T) was not determined. The IOM concluded that Agent Orange could not be linked to leukemia, bone cancer or birth defects, and was probably not implicated in cases of skin cancer, gastrointestinal cancers, bladder cancer or brain tumor.

Hidden dangers

Bleached paper

Chlorine was once commonly used to bleach wood pulp to make clean white paper. While it did this very well, it also reacted with molecules in the defoaming agents that were used, and formed dioxins. The paper industry also used pentachlorophenol as a fungicide to prevent darkening of the wood pulp by fungi, and this was also contaminated with dioxins. Both these sources of contamination have since been eliminated, but back in 1985, when investigators analysed the effluent and sludge from paper mills, they discovered dioxins. Not only that, but paper was also found to have measurable amounts of dioxin, and this naturally triggered public alarm because some types of paper, such as toilet paper and handkerchiefs, come into intimate contact with the human body.

In 1988 the EPA in conjunction with the American Paper Institute began a survey of over one hundred US plants that used chlorine to bleach wood pulp. They discovered that the pulp from which paper is made had between 3 and 6 parts per trillion (ppt) of dioxins, the sludge had 17 ppt and the wastewater from the mills had 0.024 ppt. This dioxin was coming from the wood pulp itself, which contains molecular components that will react with chlorine to form dioxins. For all the years that chlorine had been used to bleach wood pulp there must have been dioxins in paper and paper products, but the amounts had been well below the limits of detection, and only the advances in chemical analysis of the 1980s put bleached paper in the frame.

Women were worried about dioxin contamination of the materials used to make tampons, and this led to publications such as *The Sanitary Protection Scandal* from the Women's Environmental Network. Parents too were alarmed; they were using disposable nappies (diapers) on their babies, and the milk their children drank often came in paper cartons. There is no evidence that a child was ever affected. It has been estimated that six nappy changes a day would only expose a baby to 0.01 picograms (trillionths of a gram) of dioxins, whereas it would get about 100 picogrammes from its mother's milk.

The pulp industry began to look for alternatives to chlorine, and many mills now employ bleaches such as ozone, hydrogen peroxide or chlorine dioxide. Chlorine dioxide sounds as though it might also behave like chlorine itself, but this compound is an oxygenating chemical not a chlorinating chemical. The levels of dioxins in effluents and products decreases markedly when chlorine dioxide is used in place of chlorine, and this is likely to be the bleaching agent of choice when the hundreds of pulp and paper mills in the US change their processes to meet the EPA draft regulations that were published in November 1993. When these come into force in 1998 the aim will be to cut the discharge of TCDD from these mills from an estimated 350 grams of dioxins per year at present to 30 grams per year.

Dioxins in food

Lots of foods contain small amounts of dioxins. Even human milk has them, and breast milk generally has more than cow's milk. We are exposed to dioxins from the moment we are born, and we all have detectable amounts in our bodies. Most of the dioxins which humans take in come from our diet, and the main sources are meat and dairy products. Farm animals absorb dioxins from the grass and plants they eat. The dioxins are there as surface contaminants. The World Health Organisation (WHO) defines a Tolerable Daily Intake of dioxins of 10 picogrammes of dioxin per kilogram body weight per day, which is 700 picogrammes for the average human. The actual intake is about 100 picogrammes.

Other species pick up much more dioxins than humans, and fish appear to be able to concentrate them into their bodies. Catches of bottom-feeding fish like carp from Saginaw Bay, USA, have been found to have up to 700 parts per trillion of TCDD in their flesh. Most other fish have levels of around 10 ppt. Herring and salmon from the Baltic Sea in Northern Europe also have detectable amounts of TCDD, even though there are no industrial sources of dioxins in the region. Probably all meat and fish have detectable amounts of TCDD at a level of a few parts per trillion, and this we must assume is the natural background level of dioxins in everything on this planet.

Combustion

In 1977 it was discovered that burning wood produces some dioxins. Clearly anything that is burnt which is a mixture of carbon, oxygen, hydrogen and chlorine has the potential to release dioxins, even though the chemical reaction which forms them rarely occurs. Whenever we burn wood or straw, or cremate bodies, we produce some dioxins as part of the natural chemical processes that flame promotes. Timber that has been treated with chlorophenols to preserve it will of course evolve more dioxins than burning ordinary untreated wood, but now that such wood preservatives are no longer used, this problem will diminish. Every time you strike a match or have a bonfire you produce dioxins. Forest fires contaminate the whole planet with them. These we can regard as natural sources, but there are still other ways that dioxins are formed by burning, and attention has focused on one particular source, the incineration of municipal waste in cities.

Dioxin levels are highest in cities. They have been detected in car exhausts and spent motor oils, especially from vehicles which use leaded petrol. The dioxins arise because of the need to add organochlorines to the petrol to help remove the lead from the engine. When all cars ran on leaded petrol they may well have been adding thousands of grams of dioxins to the environment every year. As the use of leaded petrol has declined, so has the amount of dioxins from cars.

A much larger source of dioxin in cities has been municipal incinerators which have become the favoured method of disposing of domestic waste in many countries. In 1977 dioxins were discovered in the fly ash and flue gas of municipal incinerators in Holland and Switzerland. In Europe there are now more than 500 municipal incinerators and in Denmark and Switzerland almost all waste is disposed of this way. A modern incinerator burns about 750 000 tonnes of waste a year, generating electricity and providing low cost heat for homes and offices. Such an incinerator saves the burning of 300 000 tonnes of fossil fuel.

Modern incinerators have been greatly improved and tight standards are set for their emissions. Flue gases can be passed through carbon filters which reduces the amount of dioxin released to the environment to trillionths of a gram of dioxin per cubic metre of gas emitted. An incinerator may in fact destroy more dioxin in the waste being burnt than it produces in the furnace. According to Professor K. Nilsson, of Lund University in Sweden, if the same amount of waste were to burn in an uncontrolled landfill fire then there would be a far higher emission of dioxins. But even from an efficient incinerator some dioxin escapes, a milligram for every tonne of waste burned. This does not seem a lot, but would be 750 grams per year for a large incinerator. Environmental pressure groups have alerted those who live near incinerators to the risks, and there is often much opposition to new incinerators being built. Yet the amount of dioxin that residents would be exposed to each day would be only a thousandth of the tolerable daily intake.

Hazardous waste incinerators seem more threatening in that they are there to dispose of bulk chemicals, many of which are organochlorines, such as **polychlorinated biphenyls** (**PCB**s). These were manufactured all over the world in hundreds of thousands of tonnes per year from the 1950s to the late 1970s, and were used in electrical transformers since they were very stable and acted as insulating fluid. Other uses included hydraulic fluids, lubricants, inks, plasticisers and adhesives. Companies that specialise in incinerating chemical waste have come in for criticism, and although their plants are generally located in remote areas there have been vigorous campaigns by farmers and local residents to get them closed down. These incinerators do emit dioxins, but the amounts are relatively small because the through-put of such plants is measured in tens of thousands of tonnes per year rather than the millions of tonnes in municipal incinerators. Hazardous waste incinerators add very little dioxin to the global burden.

Nevertheless, combustion is still the main source of dioxins from human activity. For example, in 1988 the emissions of dioxins in Sweden, where environmental monitoring is well advanced, were judged to be:

Emissions of dioxins, Sweden 1988

Source	Emission* (g per year)
Incineration	110
Iron and steel industry	100
Paper industry	35
Hazardous waste	6
Motor vehicles	5

*Upper limit for all dioxins, but calculated in terms of their TEQ (see p. 180).

Although we now realise that dioxins are not the threat they once appeared to be, there is no point in allowing any to escape to the environment if it can be prevented. We must keep improving incinerators, and replace older ones as quickly as possible. Cutting emission of dioxins is more of an engineering problem than a chemical one. Many combustible materials are likely to produce a trace of dioxins given the right conditions. Such conditions may pertain in the waste gases from any burning organic material, when the speed at which these cool and the presence of solid particles may allow tiny amounts of dioxin to form. Combustion research has shown that it is not so much the air supply or the temperature of incineration that leads to dioxins being formed, but the time taken for the burning process. If this is optimised then the amount of dioxins is minimised. As a final precaution we can clean up the flue gases before they are vented to the atmosphere.

The verdict

What are we to make of all this? Throughout the 1980s environmentalists became concerned as they discovered how toxic and hazardous dioxins were, and especially when they discovered that these were present in several commonly used chemicals and household products. Thanks to these pressure groups, industry was made aware that they had to do more than just prevent accidents in their plants. Today we can look back at what has been done and realise that although dioxins still represent a risk, they are not as threatening as they first appeared. The natural environment has always had a background level of these chemicals, and we humans have them naturally within our bodies. It would appear that we have learned to live with a certain amount of dioxins. We may even have evolved to be immune to low levels of dioxins, and this may be why the expected upsurge in cancer cases resulting from industrial accidents has not materialised, even though tests on animals suggested that dioxins were highly carcinogenic.

Why then have so few people died of dioxin poisoning? Some claim there have been no deaths at all, but there is evidence that a few men who were heavily exposed to TCDD in industrial incidents, and who died a few months later, were directly affected by it. They were very unlucky. Although TCDD is deadly to animals, most humans who have been exposed to large amounts, and enough to give them chloracne, have not had their general health seriously impaired. People do react idiosyncratically to dioxins—some people break out in chloracne when exposed to tiny doses, while others can cope with much larger amounts without a single spot appearing. This was one of the more puzzling findings following the Seveso accident.

The early industrial accidents that involved dioxins were not sufficiently alarming to cause companies to invest effort or money to ensure that dioxins, and especially TCDD and TCDF, were eliminated from their products. This lack of foresight by the chemicals industry was a costly mistake. It allowed those who fought to eliminate dioxins to occupy the moral high ground, from which they have continued to direct a vigorous campaign against dioxins, even extending this to all organochlorines.

In 1993 the environmental pressure group Greenpeace issued a report called *Dioxin Factories* in which it accused PVC manufacturers in Europe of producing 'huge amounts' of dioxins. The report calculated that enough dioxins were being produced from this one area alone to supply every living soul on the planet with more than the maximum annual dose as laid down by the World Health Organisation. Now this is true, but it needs putting in context. What is the tolerable daily intake of dioxins?

To answer this question the Chlorine Institute Inc. of Washington DC, the Environmental Protection Agency and the Food and Drug Administration co-sponsored a conference in October 1990 at the Banbury Center, Cold Spring Harbor, Long Island, New York, to which they invited 38 scientists from all over the world, and especially those from environmental, biological and medical research institutes. The object of the conference was to reach an agreed figure for the tolerable daily intake of dioxins, which in some countries was as high as 10 picograms per kilogram of body weight, and in others as low as 0.0062 picograms, for example in the USA. No agreement was reached, although it was agreed that the higher limits were too high and the lower limits too low. The upper limit appears to be around 5 picograms per kilogram of body weight per day. Some countries permit less than this, such as Germany with 1 picogram, and some allow more, such as Canada with 10 picograms. The World Health Organisation also fixes the tolerable daily intake at 10 picograms.

The USA puts the tolerable daily intake of dioxins as 0.0062 picograms per kilogram of body weight, which is 0.434 picograms for the average 70 kg person. This is rather an unrealistically low limit considering the average person probably takes in between 1 and 3 picograms per kilogram body weight per day.[9] Just how tiny a picogram is can be seen by taking a gram of dioxin and expressing this in terms of an allowable daily intake. A gram would be enough to give every one of the world's 5.5 billion people their tolerable daily intake

---

[9] The average person probably has around 5000 picograms of dioxins in their body. Or put another way, this is 0.005 millionths of a gram.

for every day for a year. (For those in the USA it would supply their tolerable daily intake for over a hundred years.) It is more than likely that the 0.0062 pg per kg body weight per day, which the EPA and FDA recommend, was being exceeded when humans first walked the Earth over a million years ago. It was certainly exceeded the day they first started cooking. We can now see that the claim in *Dioxin Factories* would be correct if the chemicals industry emitted only a single gram of dioxins.

There are genuine concerns about dioxins, and especially about the potent ones like TCDD. The main sources of dioxins in the environment are given in the table below. In the USA the National Council for Air and Stream Improvement estimated in 1991 that the release of dioxins were as follows: general fuel combustion (for heating, industry and power generation) 26%; municipal incinerators, 26%; hospital incinerators, 16%; motor vehicles 7%; fuel–wood combustion, 7%; forest fires 7%; municipal sewage, 4%; pulp bleaching, 1%; other sources, 6%.

The major sources of dioxins

Industrial

Chemical industry	Pulp and paper industry
Smelting and refining of metals	Manufacture of flame retardants
Leakage from waste disposal sites	Recycling scrap metal
Warehouse fires	Hazardous waste incineration

Urban combustion

Municipal waste incineration	Accidental fires in buildings
Hospital waste incineration	Cremation
Combustion of sewage sludge	Car exhausts

Domestic fires

General heating	Wood burning stoves
Bonfires	Cigarette smoking
Accidental fires in homes	

Environmental

Forest fires	Wood burning stoves
Use of sewage sludge as fertiliser*	

---

*The dioxins in sewage sludge may come from a variety of sources.

Not included in the above table are the dioxins found in the solvents used in dry-cleaning establishments. These solvents can be organochlorine compounds which might lead you to expect that they would be contaminated with traces of dioxins when they are manufactured. This is not so—they come dioxin-free. In fact the dioxins are found only in the used solvents and they are picked up from the clothes being cleaned. These are dioxins from car exhausts, crematoria, incinerators, bonfires, and all the other domestic and urban sources mentioned in the table opposite.

The manufacture of chlorophenols and chlorobenzenes was once the main source of dioxins, but these have now largely been phased out in the West, although in some Third World countries such as India their production is just starting up. At various times they were used as wood preservatives, pesticides and weedkillers, and the annual production of chlorophenols was probably in excess of 150 000 tonnes a year. How much of the toxic dioxins they inadvertently produced is now impossible to know.[10]

Conclusions

Given enough time all dioxins spread throughout the environment because they are slightly volatile. Consequently every mouthful of food we eat contains some dioxin, including TCDD. Dioxins are almost insoluble in water but they are slightly soluble in organic matter such as fats and oils. Levels in fruit, vegetables and eggs are extremely low, and in milk are very low. In meat, fats and oils they are slightly higher, and in fish and fish oils higher still, but all the amounts are still tiny. In the human body dioxins collect in fatty tissue, and this is also the reason why human milk contains dioxins in its cream.

Although they were unaware of it, humans began to experience dioxins at first hand when they first used fire. Indeed, we may well have built up a certain degree of tolerance to dioxins because humans have been in close contact with them for so long. Hundreds of thousands of years of cooking and keeping warm with wood fires has produced a human tolerance to these chemicals which is far higher than in other species. Wood has all the elements necessary

---

[10] Based on the Agent Orange contamination figure of 2 ppm, the annual industrial production of dioxins would have been about 300 kg, making a total of 7.5 tonnes over a period of 25 years.

for dioxin formation, including up to 0.2% of chlorine as chloride. Burn wood and a measurable amount of dioxins will be produced, and they form remarkably easily, as research was to demonstrate in the following experiment. Benzene and ferric chloride were absorbed on to a silicate material and heated at 150 °C when detectable amounts of dioxins were generated. This sounds like a rather odd chemical reaction, but it mimics a primitive wood fire on solid earth. The surprising thing was that dioxins were generated at relatively low heat. Here we are talking about a range of dioxins, not just TCDD, and this is how it is with wood fires.

In 1985 Hermann Poiger of the Swiss Federal Institute of Technology was so sure that dioxins were fairly harmless to humans that he ate some in order to chart their progress through his own body. He found they concentrated in his fatty issue and had a half-life in his body of five years. Of course, he took only a tiny dose.

In 1981 the amount of dioxins released to the global environment by human activity was estimated to be about 1500 kg from the chemicals industry, about 13 kg from combustion (fuels and incinerators), 0.2 kg from motor vehicles, and a mere 0.002 kg from cigarette smoke. Since then the levels of dioxins have been falling in the West as the major industrial polluter, the chemicals industry, has controlled the making of the chlorophenols, 2,4,5-T, 2,4-D and PCBs, and in some countries their manufacture has ceased altogether.

Less well publicised is that Nature itself makes chemicals like these. The soil fungus *Penicillium* produces 2,4-dichlorophenol. Grasshoppers make 2,5-dichlorophenol, and the Lone Star tick (*Amblyomma americanium*) uses 2,6-dichlorophenol as its sex attractant. Nature has ways of getting rid of chlorophenols, which are broken down in soils by the white rot fungus *Phanerochaete chrysosporium* which can even digest DDT, PCBs and pentachlorophenol. PCBs are also destroyed by the soil bacterium *Pseudomonas*.

So what is the position today with regard to dioxins? Should we still feel threatened by them? The answer is 'yes' if we are talking about TCDD itself and the other highly toxic dioxins. We must remain alert to any concentration of these compounds contaminating the environment as they did as Seveso. The answer is 'no' if we are talking about the general background levels of dioxins of the kind which we encounter when we light a cigarette, have a bonfire in the garden, drink a glass of milk, or pass an incinerator. Among these dioxins there is some TCDD and TCDF, but as this chapter has revealed, most of them will be the harmless kind and so none of these activities puts us at risk.

Professor Christoffer Rappe, the world's leading researcher into dioxins, has been quoted as saying: 'More people make their living from dioxins than ever suffer from them.' He did not mean just the manufacturing chemists, or the toxicologists and analysts. He was referring to people such as medics and lawyers who have to deal with people who think they have been affected by exposure to dioxins and wish to take legal action. Nor must we overlook those who made political capital out of dioxins. In Holland the Green Party used the news that cow's milk was contaminated with dioxins to score impressive gains in elections.

In 1991 Michael Gough of the Center for Risk Management, Washington DC, reported on the human health effects of dioxins in a paper in the journal *The Science of the Total Environment*. His findings were rather reassuring, and he concluded: 'no human illness, other than the skin disease chloracne, which has occurred only in highly exposed people, has been convincingly associated with dioxins.'

To end this chapter I would like to quote Professor W. Gribble of the University of Hawaii at Manoa, Honolulu. In response to the oft-repeated claim by environmental campaigners that organochlorines do not occur in Nature, he wrote a 40-page review in the October 1992 edition of the *Journal of Natural Products* in which he discussed hundreds of known examples of organochlorines which are made by living things. He also concluded of dioxins: 'There is no scientific evidence that dioxin (TCDD) causes any serious health effects in humans apart from the skin disease chloracne and some reversible liver dysfunction. Nevertheless the extraordinary toxicity of some polychlorinated dioxins (PCDDs) and related compounds, such as the polychlorinated dibenzofurans (PCDFs) in some animals is reason enough for the continued study of these compounds, especially since it is now recognized that PCDDs and PCDFs are in fact, natural products.'

As far as our animal friends are concerned we should continue to research the dioxins, but as far as humans are concerned the dioxin scare is over.

CHAPTER EIGHT

Nitrate

The nitrogen cycle in Nature; food; fertilisers; nitrate in natural waters; nitrate in the diet; nitrate and human health; conclusions

IN the previous chapter we considered the dioxins, toxic chemicals that we unknowingly dissipated into the environment. Thankfully they are less hazardous to humans than tests on animals would indicate. In this chapter I want to talk about another chemical which worries people, yet one that we manufacture on a massive scale, and deliberately spread far and wide. This chemical is the fertiliser **nitrate**,[1] which many people believe damages the land, pollutes rivers, contaminates water supplies, and threatens human health. Nitrate has been guilty of all these charges at some time or other. Some blamed overproduction by industry and overuse by farmers, because it appeared that as nitrate use went up, so did the level of nitrate in rivers and drinking water. The visible evidence of excess nitrate in natural waters was the growth of weeds and slimy algae; the less obvious effects were said to be sick babies and cancer.

Some groups see nitrate as a threat from the moment it is produced in a chemical plant, to the time it ends up in our stomach, and all stages in between. Others would like to see an end to inorganic nitrate fertiliser production, and a return to organic manures to replenish

---

[1] A reminder that when the name of a substance appears in **bold** type, it means that there is more about its chemistry in the Appendix.

farm land, claiming that we could break this chain of pollution and so save lives. In fact the opposite would happen—if we were now to stop making nitrate fertilisers *millions* of people would starve to death. Nor is the link between fertiliser production and environmental pollution as simple as some people assume. Indeed, there may be hardly any connection between them.

Little that is said about nitrate is true, much is distorted, and explanations of how it affects our health are based more on fear than on fact. We need nitrate to grow enough food to feed the people of the world. The effect of inorganic nitrates on food production is well illustrated by looking at the history of Britain. By 1900 the British Empire was the greatest the world has ever known, but in the two World Wars in the first half of this century it discovered its Achilles' heel: the heart of the Empire was an overcrowded island of 50 million people which had to import most of its food. This vulnerability was not lost on the enemy whose submarines sent ship after ship of food to the bottom of the Atlantic. 'Dig for victory!' was the call to which millions responded, but despite all the digging, and the drafting of young women to work on farms, Britain still grew less than a third of the food it needed.

Fifty years after the Second World War, and with a few more million mouths to feed, the farmers of that overcrowded island grow twice as much of their staple crops, and on less agricultural land. Indeed, the British Isles could be self-sufficient in food if it so wished, but people now put land to other uses. Fields that once grew wheat and potatoes are planted with trees; meadows on which cows and sheep grazed have become theme parks and golf courses; and farmers whose land once bloomed yellow with oilseed rape are now paid to let it lie fallow.

Such is the abundance of food grown in developed countries that some farmers are returning to the methods of a bygone age, or 'organic' farming as they prefer to call it. They plant crops in strict rotation, and they fertilise with manure and compost generated on the farm. If all farmers were to do this, countries like the British Isles would again have to import food on a massive scale. The bounty of modern agriculture is only possible because of chemical fertilisers, and in particular nitrate. To understand why this is so we need to understand the environmental chemistry of **nitrogen**.

The nitrogen cycle in Nature

Nitrogen is essential to life because it is one of the five elements which make up DNA; the others are hydrogen, oxygen, carbon and phosphorus. Apart from its role in DNA, nitrogen also has a part to play in hundreds of other molecules, and especially in **amino acids** which are the building blocks of protein. The average human being weighs 70 kg (155 pounds) and contains about 1.8 kg (4 pounds) of nitrogen, mainly in the form of protein.

We could trace the route by which an individual nitrogen atom in DNA or protein started its journey from the nitrogen gas of the air and passed through a series of different forms in plants and animals before it ended up as part of our body. We could also trace its journey back again. When material of which it is part breaks down, the nitrogen atom may be recycled within out body, or it may be excreted in our urine as amino acids, **ammonia**, **urea** or other chemicals. But what is one animal's waste is another organism's food, and this discarded nitrogen can be used by bacteria or plants as their sources of nitrogen. A particular atom of nitrogen may circulate round the food supply many times before it meets an organism that turns it back to nitrogen gas and expels it to the air. Then the cycle is complete. The term 'nitrogen cycle' refers to the movement of nitrogen through the environment, and the illustration opposite shows the main paths around which it revolves.

The atmosphere is 79% nitrogen gas, N_2, but this is useless as a fertiliser resource for plants because it is a very unreactive molecule, and though it will dissolve in water to a limited extent, it still does not undergo chemical reactions. Generally high temperatures are needed to get N_2 to react with oxygen. **Nitric oxide** (NO) and **nitrogen dioxide** (NO_2) are produced naturally by lightning and by combustion of fuels, and these can be washed to earth.[2] To be of use they have to be turned into nitrate (NO_3^-) or ammonia (NH_3) which plants can use as nutrients.

Reaction of nitrogen with hydrogen to form ammonia is rather subtle when carried out in Nature. There are micro-organisms that can take nitrogen gas and add hydrogen atoms to it at outdoor

---

[2] These are often referred to as NO_x, where the x stands for one or two oxygen atoms. But the main one is NO_2.

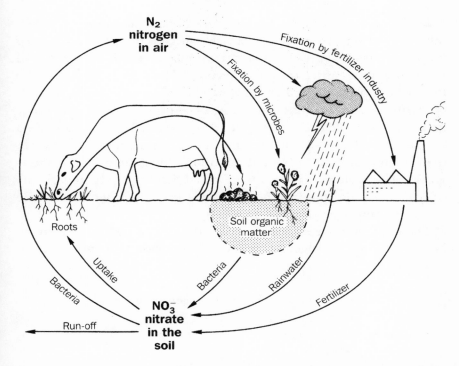

The nitrogen cycle in nature. The main 'reservoirs' of nitrogen are nitrogen gas in the atmosphere and nitrate in the soil. In addition to these inorganic forms there are thousands of organic nitrogen compounds which are part of living things.

temperatures, and the most important ones are the so-called nitrogen-fixing bacteria. Some can do this with apparent ease, such as the free-living *Clostridium pasteurianum*, which is anaerobic—in other words it can live without oxygen from the air—and *Azotobacter vinelandii* which is aerobic, in other words it needs air. The most important nitrogen-fixing bacteria are those which live in symbiotic relationships with plants, such as *Rhizobium* which lives in nodules on the roots of legumes such as peas and beans. The plants provide the bacteria with carbohydrates while the bacteria provide the plants with ammonia. These bacteria are the first step in the nitrogen cycle on land, and there are others that begin the food chain of the sea, such as *Cyanobacteria* and *Rhodospirillum* which fix nitrogen gas dissolved in sea water.

In addition to the bacteria which turn N_2 into ammonia, there are others called nitrifying bacteria which can take the process one step further: *Nitrosomonas* can convert ammonia to **nitrite** (NO_2^-), and *Nitrobacter* turns nitrite to nitrate (NO_3^-). This may be absorbed by the roots of plants, which in turn may be eaten by animals. Animals can return a lot of the nitrogen to the soil as urine or dung, and it may go round the biological cycle again. Alternatively it may be absorbed by bacteria which use nitrate (and *nitrite*[3]) as a source of oxygen, thereby converting it to N_2 and returning it once more to the atmosphere. The bacteria that do this are called denitrifying bacteria, some of which can also produce nitrous oxide gas, N_2O, which likewise returns the nitrogen to the atmosphere.

The nitrogen cycle in Nature is now dominated by human activity. Nature and humans together fix about 275 million tonnes of nitrogen annually in various ways:

Natural fixation of nitrogen (175 million tonnes annually):

agricultural crops	80
wood and forests	40
marine plants	35
lightning	10
other	10

Nitrogen fixation due to human activity (100 million tonnes annually):

fertilisers	70
combustion	20
other	10

The amount fixed by plants grown for human consumption is the single largest contributor, and this is achieved by crops such as legumes (peas, beans, etc.) and rice. If this is seen as part of the nitrogen fixed by human endeavour, then we are responsible for fixing 180 million tonnes a year, far exceeding Nature's 95 million tonnes. The most important part of the nitrogen cycle is that in the soil. An instructive way of thinking about this is to consider all the nitrogen compounds

---

[3] The name nitrite is very like nitrate, and yet in chemical terms they behave very differently. We need to keep this in mind for later in the chapter. To avoid confusion *nitrite* will be in italics.

in the soil as a nitrogen reservoir, which is added to and taken from each year. The total pool is a mixture of many forms, from simple nitrate ions dissolved in the water of the soil, to complex proteins and molecules in the cells of microbes and all the many other creatures living there.

We add to this reservoir when we fertilise the land, and both inorganic ('artificial') fertilisers and organic (manure and compost) fertilisers provide nitrogen compounds. We tap this reservoir when we grow and harvest crops, because plants use their roots to absorb nitrate from the water of the soil.

Food

We get a lot of the amino acids we need by consuming the protein of other living organisms, such as the flesh of animals, or by raiding the protein stores of plants, such as their seeds. The world relies on a small number of crops that we have cosseted for generations. The top four are wheat (annual production about 400 million tonnes), maize (350), rice (350), and potatoes (300). Others in the big league, with 100 million tonnes or more, are barley, sweet potato, cassava and soyabean; and sugarcane and sugar beet together yield 120 million tonnes of sugar a year.

Some of these crops are excellent sources of protein. One such is the soyabean, which has been popular in China for thousands of years, and is now among the top four crops grown in some Western countries such as the USA. Soya flour, which is very high in protein, is increasingly used in bread and other products. The components of a soya bean are protein (42%), carbohydrate (29%), and vegetable oil (21%). Soya flour is 50% protein, and this is much higher than lean beef (20%), wheat (14%), eggs (13%) or milk (4%).

A well fertilised hectare[4] of soyabeans will yield 500 kg of protein, ten times as much as we could get from cattle grazing on the same area of grass. Most of the nitrogen that passes through an animal in the course of its life is returned to the land, and this tends to be distributed rather unevenly, so contributing to the problems of leaching, as we

---

[4] A hectare is the area of a square 100 metres by 100 metres, i.e. 10 000 square metres. In Imperial units this is about 2.47 acres (an acre is 4840 square yards).

shall see. Protein is about 15% nitrogen, so a crop of soya beans will remove 75 kg of nitrogen per hectare from the land, whereas cattle taken to slaughter will remove only 6 kg. We can see why the former is not a sustainable crop without a large input of nitrogen fertiliser, but a herd of cattle can be—at least in theory.

The value of most crops lies in the vegetable oil or carbohydrate they produce, neither of which contains nitrogen; but without access to nitrogen such crops would not grow. All plants need nitrogen to be able to carry out photosynthesis, which is the process that converts carbon dioxide from the air to carbohydrate. This reaction requires sunlight and takes place in leaves. In order to intercept sunlight efficiently, a hectare of crop needs to maintain a green canopy of three hectares of leaf area. Cereals need about 90 kg of nitrogen to form such an amount of leaf.

This, then, is why farmers apply nitrate fertiliser at the start of the growing season. It would be little use applying it in late summer, and if they were foolish enough to do so they would lose most of it in the rains of the following winter. The nitrogen of the leaves eventually ends up in other parts of the plant, such as stems, roots and storage organs, which may be tubers or grains. The proportion of nitrogen that is removed at harvest time will vary from as much as 75% in the case of cereals, to as little as 35% for sugar beet. What is left behind generally adds to the organic reservoir in the soil, the so-called humus.

The most important source of nitrate in soil is the humus. A few percent of the nitrogen in this is converted to nitrate each year, and this is taken up by plants in the spring and early summer when it is needed for growth. A lot of the nitrogen from humus is released too late in the year to be of use to plants, and it may be leached from the soil by rain in the autumn and winter. The reservoir of nitrogen compounds in soil is very large, and only a small part is accessible to crops. It is possible to tap this reservoir and grow crops without fertilisers year after year, although with slowly diminishing returns.

If we did nothing to refill the nitrogen reservoir in the soil, but simply grew crops and harvested them, we would find yields falling as the humus store was used up. Eventually the land would sustain only the amount of vegetation that could be supported by the nitrogen that Nature provides each year. Under such conditions the Earth might just be able to feed the current population of 5.5 billion people if we cultivated as much land as possible. Today about 1.5 billion hectares (10% of the global total) are cultivated, but a further 2 billion hectares, mainly in South America and Africa, could be brought into production. Yet even this would not be enough because the human population would continue to expand, so we would have to cultivate yet more land. The alternative is to farm existing land more intensively. This is only possible if we replenish the nitrogen pool in other ways, and the best way to do this is to fertilise the soil with inorganic nitrates. Yet nitrate fertilisers are said by some to be a major environmental threat, despite the fact that millions of us are already dependent on them for the food we eat.

The Earth and its organic nitrogen resources were outstripped by our expanding population many decades ago. The gap has to be filled with inorganic fertilisers. The Worldwatch Institute reported in *State of the World 1990* that the Earth's food output would plummet by 40% if the use of these fertilisers were to be discontinued. In other words about 2 billion people are now alive, and are kept alive, by the nitrate which comes from fertiliser factories.

Fertilisers

'And he gave it for his opinion, that whoever could make two ears of corn or two blades of grass to grow upon a spot of ground where only one grew before, would deserve better of mankind, and do more essential service to his country than the whole race of politicians put together.'

So wrote Jonathan Swift in *Gulliver's Travels*, published in 1726. Two hundred and fifty years later we have done just this over large parts of the globe. Indeed, farmers can now grow not two, but four and even six ears of corn where only one grew before. The miracle has been achieved partly by breeding better strains of corn, but mainly by applying fertilisers, and especially nitrogen fertilisers, to the soil. The value of fertilisers can be illustrated by looking at the world's longest running agricultural research station.

The 150-year experiment

About 25 miles north-west of London is the Rothamsted agricultural research centre. It was set up in 1843 by John Lawes and Henry

Gilbert. Lawes manufactured fertiliser phosphates and Gilbert was a professionally trained chemist. From its inception Rothamsted's brief was to examine the effects of artificial and natural fertilisers, and this it has continued to do; but it has also become a major centre of fundamental research into all aspects of agriculture, from recycling waste within peasant economies to genetic studies on ways to give plants an inbuilt resistance to predators.

What fascinate most visitors to Rothamsted are the so-called Classical Experiments, which are plots of wheat, grass, and other crops that have been grown on the same strips of ground year after year for over a century, with the same amounts of inorganic and/or organic fertilisers added to each plot every year. Rothamsted even has a herd of cows which are fed in the traditional manner so that their manure can be used as organic fertiliser. They also have a plot of land which has been growing winter wheat for 150 years with no input of fertiliser at all. Every year the soil is sown with wheat, the crop is allowed to grow with nothing being added to the soil, and then it is harvested. Nothing else is done to the land until the next season when it is ploughed and sown again.

When you compare the yields of crops from these experiments, you can see the benefit of fertilising the land. The yields of wheat and potatoes per hectare, averaged over many years, are given in the table below.

The Classical Rothamsted fertiliser field tests

Fertilisers	Crop yields*	
	Wheat	Potatoes
none	1.2	8.2
NPK** (1:1:1)	5.5	20.4
farmyard manure	5.8	32.9
NPK** (5:1:1)	6.8	35.6
farmyard manure + N	7.7	37.8

*Tonnes per hectare   **N = nitrogen, P = phosphorus, K = potassium

What first strikes us about these figures is that the unfertilised strip of land can still produce an average crop of 1.2 tonnes of wheat. It will even yield 2 tonnes per hectare in a good year, although only 0.5 tonne in a poor year. A neighbouring plot to which nitrogen fertiliser has been added every year has never produced less than 2 tonnes even in a bad year.

In the late 1960s new varieties of wheat were introduced at Rothamsted, and these could better use the nitrogen in the soil. On the fertilised plots yields went up to 7 tonnes, while it stayed at around 1 tonne on the unfertilised soils.

Growing a crop and taking it to market means that we are removing nutrients from the land, and so it is with the Rothamsted unfertilised plot. The wheat cropped from this plot removes about 12 kg of nitrogen, 5.5 kg of phosphorus and 7 kg of potassium; most nitrogen and phosphorus are in the grain, and most potassium is in the straw. Where do these nutrients come from? Indeed, what sustains the fertility of this soil? The wheat needs nitrogen, phosphorus, sulfur, potassium, sodium and a few other essential minerals. It must get these from its roots, which absorb them from the water in the soil.

In the Middle Ages, when crop yields began to fall, a field was left to lie fallow for a year. This had the effect of allowing weathering to break down soil particles and release phosphate and potassium, and also to add more nitrogen by way of rain water, natural fixation and bacterial release from humus. The immediate beneficiaries of this bounty are the microbes in the soil, of which there are vast amounts, and these act as the real reservoir of nutrients which they release when they die and decay. It is this organic reserve from which plants derive their nutrients in undisturbed soil. Ploughing, sowing, harvesting and removing crops upsets this natural balance. With no replenishment the nitrogen reservoir dries up, and yields fall until the vegetation is sustained only by rainfall and bacterial fixation, and we reach the minimum of the Classical Experiment.

The table on the next page shows how much fertiliser is used worldwide, and the figures include phosphorus and potassium as well as nitrogen. The amounts of these fertilisers manufactured each year are roughly 70 million tonnes of nitrogen (mainly as nitrate), 35 million tonnes of phosphorus (as phosphate), and 30 million tonnes of potassium (as the chloride). Nitrogen is the key nutrient because it is the one most likely to be in short supply. It needs replenishing yearly, and if we do so the effects are dramatic. In Europe, for example, the amount of nitrogen used in fertilisers has increased from less than four million tonnes to nearly ten million tonnes in the last 25 years. As the

Changes in the use of fertilisers throughout the world, and major crops produced, from the early 1950s to the early 1980s

	North America	South America	Europe	Asia	Africa	Russia
Fertilisers						
1950s	6	0.3	9	2	0.5	2
1980s	24	4	32	29	4	19
*Crops**						
1950s	149	26	294	768	20	141
1980s	504	94	623	1402	53	373

*Millions of tonnes per year. Averages have been taken for the early 1950s and early 1980s.
**Millions of tonnes per year. These are the totals for the major staple foods: wheat, rice, potatoes, barley and maize.

amount has gone up so have crop yields, and wheat production has doubled from 2.5 tonnes per hectare to 5 tonnes. Maize yields went up to 6.5 tonnes.

The amount of fertiliser used per head of population varies widely. For example in Europe and North America it is around 30 kg per person per year, while in Africa it is less than 5 kg. Wherever modern fertilisers are used crop yields have gone up, sometimes increasing threefold, as the table shows. Of course not all the increase is due to fertilisers, and in developing countries there has been an increase in the land brought into cultivation, as well as an increase in yields from new varieties of crops.

Inorganic fertilisers have affected farming in four ways:

1. They have permitted naturally infertile soils to be brought up to a level of nutrition that will support the growing of staple crops. For example, in Australia cereals are now grown on 16 million hectares compared to 6 million thirty years ago.

2. They have eliminated the need to replenish the nitrogen in the soil by older methods such as leaving the land fallow for a year, or by planting legumes or clover which can fix nitrogen in their roots, and then ploughing these back in. The sequence of the traditional Norfolk system of annual crop rotation is: wheat → root vegetables → barley → clover → and then back to wheat. This system alternated human and animal crops, and worked very well. We still

need to rotate crops, but today the reason is pest control rather than to maintain soil fertility.

3. Fertilisers increase yields on farmed land, thereby allowing other land to be used for different purposes, such as forestry or recreation, and some may even be left undisturbed as a haven for wildlife. Of course intensive farming also frees land for less desirable uses such as sprawling suburbs, supermarket car parks, and six-lane motorways.

4. Fertilisers influence which crops are grown because they affect their profitability. Farmers will tend to cultivate those crops which respond best to fertilisers. The cost of fertilisers finally ends up in the price of the crop, but how much it contributes is difficult to calculate because of the many fixed costs in farming. Happily, inorganic fertilisers are cheap, although some see this as a disadvantage, encouraging profligate use and causing pollution.

Organic or inorganic nitrogen?

There are some who think that inorganic fertilisers are alien to the environment, and they advocate 'organic' farming with 'organic' fertilisers as the natural way to replenish the land. Organic farming is a luxury that only advanced economies can afford because it is highly inefficient. Organic farming means less produce and higher prices, but some city dwellers are prepared to pay a premium for such food in the belief that it is somehow healthier. They may even imagine that the protein it contains is made up of 'organic' nitrogen, produced from 'organic' nitrate, which is provided by the 'organic' fertilisers which organic farmers spread on their land. A little chemical knowledge shows that this is not so: there is only one kind of nitrate.

Nitrate is a simple negatively charged molecule consisting of a nitrogen atom surrounded by three oxygen atoms, and its formula is NO_3^-. Organic nitrate is also NO_3^-. Nitrates are salts of **nitric acid**, HNO_3, which is used in the chemistry laboratory and in the chemicals industry. The acid is made by reacting ammonia with oxygen, and it is then converted to salts, such as **ammonium nitrate**. Today 70 million tonnes a year of this are made industrially. This is the material referred to as inorganic nitrate, of which 90% is used as fertiliser and about 10% is used in explosives.

There are some natural nitrate minerals. Chile nitre, for example, is naturally occurring sodium nitrate, and was once mined on a large scale for the fertiliser and explosives industries. It is still mined for the relatively high level of iodine that it contains. Another long-known nitrate is saltpetre, which is potassium nitrate. This has been used for centuries as a component of gunpowder, and for centuries it came from organic sources, scraped from the walls of latrines and leached from cesspits, where it is produced by bacteria working on the nitrogen compounds in human excreta. No use was found for it until the discovery of gunpowder, after which it had a major impact on world history. In the Napoleonic Wars in the early 1800s the British Navy stopped supplies of nitrate reaching France from the teeming cities of India and the Far East. The French had then to use all kinds of rotting human and animal wastes to produce saltpetre.

For the wars of the 20th century, the demand for nitrate for explosives was far in excess of domestic production on this scale. The First World War might have ended with a quick victory for the British Empire if the navy blockade of Germany had been effective in cutting off supplies of nitrate, but it was to no avail, and that terrible war was to last four years and end in stalemate. The Germans no longer needed to import nitrate because their chemicals industry had found a way of using nitrogen from the air to make nitric acid and nitrates. They had discovered a chemical method of fixing N_2 gas, and on a scale that would one day rival that of Nature. The Haber process, named after its developer Fritz Haber (1868–1934), turns N_2 into ammonia. This can then be oxidised to nitric acid and nitrates. In 1918 Haber was awarded a Nobel Prize for his chemical achievements.

Jonathan Swift may be rejoicing in Heaven at the abundance of modern farming with nitrate fertilisers, but here on Earth we wonder if we are doing the right thing. More and more we hear the warnings of environmentalists that we should give up our inorganic fertilisers and return to the methods that farmers used when Swift was alive. Their advice is to use organic fertilisers, which they see as sustainable fertilisers—in other words, producing them requires no fossil fuels or mined resources.

At first glance there seems to be advantages in using animal manure, including human sewage, and compost made from municipal waste. The Rothamsted results in the table on p. 212 showed that farmyard manure is just as good as inorganic fertiliser, and what is more it appears to come free—provided you have farm animals to excrete it and pasture on which they can feed. Natural sources of potassium such as composted vegetation, and the traditional phosphate source, bone meal, are also part of Nature's bounty. Inorganic fertilisers on the other hand cost money because they come from chemical plants, and producing them requires energy and raw materials. Closer analysis, however, reveals that organic fertilisers are not free either: it is just that the costs are hidden. Animals that provide the manure have also to be reared and fed, and the food or grass they eat is also grown on farm land that cannot be cropped indefinitely without it too losing its fertility.

Plants are capable of absorbing nitrogen only as nitrate or ammonia, which they then convert to nitrate. Ammonium nitrate makes an ideal fertiliser, and whether it comes from the chemicals industry, or from manure, it makes no difference to the roots of a plant. Indeed, plants prefer the artificial variety because it is generally applied in spring and at just the right time to maximise their growth.

Despite what the advocates of organic fertilisers may believe, both types of fertiliser, organic and inorganic, supply the same essential nutrients. They differ in that they supply differing amounts and at different times. Organic ones have the advantage of recycling waste, and they release their nutrients over a period of months or even years as they decompose slowly in the soil. There are certain disadvantages to set against these benefits: the nutrient content of organic fertiliser is difficult to judge and may be low for certain elements. Moreover, the slow release may mean that it gives up its nitrate at a time when plants do not require it, so there is a higher risk of its nitrate being leached from the soil by winter rain.

The nitrogen in animal feed mainly returns to the land, but the nitrogen in human food rarely does so. Most exits down sewers and eventually ends up discharged into rivers and is carried to the sea. This seems a waste of a hard-won resource, and so more and more slurry from city sewage works is being taken to fertilise farm land—in other words, to go through the cycle again. It cannot replenish the reservoir by itself, and in some cases the sewage may be polluted with heavy metals from industry, making it unsuitable as a fertiliser for food crops.

A way round this problem is to limit its use to permanent woodlands, and to crops that are processed before they are used for human consumption. City dwellers produce sewage sludge all year round, whereas plants really need it only during the growing season. Like farmers who store animal slurry, we could store human slurry for spreading during these seasons, but the cost would be prohibitive. In any event there is the expense of transporting tankers of human slurry or trucks of compost over long distances.

These considerations apart, the benefits of organic fertiliser are still enough to outweigh the disadvantages. The problem with organic fertilisers is that there can never be enough of them to supply all the nitrogen that is needed for new crops, and there is no escaping that simple fact of environmental chemistry. No matter what we do, we cannot defeat the net transfer of nitrogen from countryside to city, for which no amount of manure, be it animal or human, can compensate.

Organic farming also benefits from another source of nitrogen fertiliser, albeit an inorganic one: the dissolved nitrogen oxides in rainwater. You might be surprised to learn that most of the nitrogen in rainwater is not there as a result of lightning, but comes from car exhausts and industrial pollution. Today's rainwater brings down more fixed nitrogen than it used to, principally because of air pollution. This bounty of nitrogen also falls on organic farms, and they may get as much as 50 kg of nitrogen per hectare from rainwater.

Inorganic fertilisers have the advantages of being easy to make, store and transport. Inorganic fertilisers can be put on the land at the time they are needed, thereby preventing wastage and nitrate run-off. They are cheap, but this is a mixed blessing in that it may encourage the misuse of land. If you look back to the table on p. 212 you will see that, in terms of crop yield, the blend which is richer in nitrogen gives the best result, and modern farming methods can improve even on these Rothamsted yields. By adding lots of ammonium nitrate fertiliser it is possible to grow over 10 tonnes of wheat and 50 tonnes of potatoes per hectare.

Generally the debate about nitrate fertilisers has been fought over which type is best: inorganic nitrate or organic nitrate. To the growing plant it matters not at all—they are chemically the same, and so is the nitrate from microbes in the soil, and that derived from rainwater.

Nitrate in natural waters

The phosphate and nitrate of artificial fertilisers have come in for a lot of adverse publicity over the years, and they were partly responsible for the eutrophication of rivers and lakes. The word eutrophication is derived from the Greek words for over-nourishment, and this happens when there is too much phosphate and nitrate entering a body of water. The species which most quickly uses these nutrients are the algae, and they multiply rapidly and choke the water with their slimy green scum. The living algae at the surface block out the sunlight for other plants, while the dead algae at the bottom use up dissolved oxygen as they rot away. Other creatures such as fish die because they are unable to breathe, and the system goes out of balance and may become a stinking mess. Nor is this only a setback for wildlife. Humans may be deprived of a source of drinking water and a place of recreation. Back in the 1950s and 1960s this happened to some of the Great Lakes of North America.

Because phosphate and nitrate are the main component of fertilisers the cause seemed obvious: streams and rivers draining water from intensively fertilised land were being over-supplied with these nutrients. Fertiliser use and eutrophication appeared to go hand in hand, and the link between them was often made by environmentalists. The chemical industry was held to be responsible for encouraging farmers to use more and more nitrate to get higher and higher yields. Of all salts, nitrates are the most soluble and the most likely to be leached from the land to pollute rivers and lakes.

In fact the effluent from sewage works was mainly to blame for eutrophication. The problem was worse than just eutrophication: nitrates were contaminating drinking waters, and this chemical was known to cause a life-threatening condition known as 'blue-baby' syndrome (see p. 228). Some people even blamed nitrate for causing cancer.

Leaching nitrate from the land

The nitrate in rivers and other natural waters comes mainly from the soil, of that there is no doubt. Nor is there any doubt that fertiliser nitrate is responsible for some of it, and sometimes for a lot of it, especially when it is washed from the land by heavy spring rains or floods. However, these are relatively rare events, and do not explain the normal levels of nitrate in natural waters. So what is the cause? There is no simple answer. Even if we can be certain from which soil the water has drained, we cannot be sure which type of fertiliser, organic or inorganic, is to blame.

Although nitrate fertilisers are used in large amounts, they do not account for the raised levels of nitrates in rivers. This comes mainly from organic sources such as soil microbes or animal manure. Environmentalists assumed that it was the fault of farmers for putting too much nitrogen fertiliser on their land. In fact it is more likely to have been caused by their forefathers spreading manure or simply ploughing up grassland. It can take decades for nitrate from soil to seep into underground water sources and find its way into rivers.

Ungrazed grassland is least likely to leach nitrate, even when inorganic fertiliser is added, provided this has been applied at the right time of year. Nitrate loss is accelerated if animals are put on the land because of their urine and dung. Urine is the main route for an animal to dispose of unwanted nitrogenous material from its body, and this tends to be deposited on a small patch of ground which is then over-fertilised. This local bounty of nitrogen exceeds plant uptake and so remains unused at the end of the season. A meadow of contented cows may look like a perfectly balanced organic system as the cows eat the grass and fertilise the soil, but it is a system out of balance. From such a field there will be a leaching of more nitrate in the autumn and winter than if the soil had been evenly fertilised in the spring with inorganic nitrates, and cattle kept off it.

The nitrate in farmed soil comes from four main sources:

1. fertilisers, i.e. inorganic nitrate

2. animal manure, i.e. organic nitrate

3. soil microbes, i.e. nitrate from microbial wastes, and

4. rainwater, which contains dissolved nitrogen oxides which are nitrate precursors.

All can contribute to nitrate loss, and this will happen if too much nitrogen fertiliser is applied in spring for the growing crop to use, or too much manure or slurry is spread in the autumn, or the microbes are disturbed by ploughing, or the rain falls through heavily polluted air from an urban area. Each contributes to the nitrate which leaches from the land.

Inorganic nitrate. The optimum amount of nitrogen to benefit a particular crop has been known for over a century. For example winter wheat needs 214 kg of nitrogen per hectare, so amounts in excess of this will simply be wasted. If it gets a little less fertiliser than this economic optimum, then the amount of residual nitrate left at the end of the season will be very small.

Computer programs have been developed to help farmers cut down the loss of nitrate from their land. Input the crop, the time of year, and the existing natural nitrogen content of the soil, and the program will calculate the right amount of nitrogen fertiliser to add to maximise crop yield and minimise waste. Of course a big unknown factor in all this is the weather, and in particular the rainfall, although irrigation can redress a dry season.

Organic nitrate. Adding manures also leads to greater yields. In the very long term, organic matter added as fertiliser increases the store of organic nitrogen in the soil, and this in its turn breaks down to nitrate at a later date, and may then be leached. Any agriculture that builds up soil fertility and organic matter will increase leaching of nitrate if it is not properly managed.

Organic manures come in the form of semi-liquid slurry from livestock that is housed, and solid matter from farmyard manure and poultry. In many countries these sources of nitrogen make up about a third of all fertilisers applied to the land. Unfortunately they are not necessarily applied at the best time for a growing crop. Indeed, a traditional time for spreading muck is the autumn, on the stubble of a previous crop, and this leads to nitrate leaching during the coming winter because there is rapid degradation of the nitrogen compounds to nitrate. Experiments have shown that pig slurry will lose a third of its nitrogen as nitrate in this way. It is possible to slow down this loss by controlling the bacteria that cause the breakdown. This can be done with **dicyandiamide**, which will suppress the microbial action that causes the conversion of the ammonia in the manure to nitrate.

Even organic farming can cause leaching of nitrate to rivers. Crop rotations in which the land is fertilised by growing a nitrogen-fixing legume crop, and then killing this by ploughing it in, can add several hundred kilograms of nitrogen to a hectare of land, but a lot of this may be leached away over winter and be gone before the next crop can use it.

Microbial nitrate.   In the top 25 cm (10 inches) of arable soil, known as the topsoil, there may be 5 tonnes (5000 kg) of nitrogen per hectare, locked up in humus and living things. Most of this is unaffected by rainwater. In undisturbed grassland and forest there may be as much as 20 tonnes (20 000 kg) of nitrogen per hectare. It can be slowly released as nitrate by the action of enzymes.

This natural fertilising action can provide 45 kg of nitrogen as nitrate per hectare per year. We can upset the natural ecosystem by ploughing it up, which stimulates the microbial activity, and this is probably the most important cause of nitrate loss from any soil. When permanent grassland is ploughed up for arable farming it releases a huge amount of nitrate, eventually up to 4 tonnes (4000 kg) per hectare. This loss is far in excess of any amounts from nitrate fertilisers. The best course of action to prevent nitrate loss is to leave old-established grassland well alone.

Rainwater nitrate.   We can also blame rain for adding to the nitrate run-off from farms. The gases dissolved in the rain, along with dust particles washed out of the air, can add up to 50 kg per hectare per year of nitrogen to the land, but most of this cannot be used by plants because it falls out of season.

Nitrate loss to rivers also depends upon the rock below the soil. If this is impervious then the water in the soil drains away into ditches or streams, but if it is porous rock, such as limestone or sandstone, then the water, and the nitrate dissolved in it, trickles slowly through and accumulates in the water table below. If we use this for drinking water by sinking a well or drilling a borehole, then we may be consuming nitrate carried there a century or more ago.

On a typical modern farm most of the useable nitrogen comes from inorganic fertilisers. The next largest amount comes from manures, then organic soil residues, and finally a small amount from rainwater. The nitrate that is leached from the soil also comes from these sources, but not in the same order. Organic soil residues provide most, then manures, then rainwater, and finally a small amount from unused inorganic fertilisers. This order is the opposite of what we might expect, and indeed it is contrary to the claims of those who condemn inorganic fertilisers. You may be wondering how we know that inorganic nitrate is the least to blame. Earlier I stressed that nitrate is nitrate, no matter what its source. There is no different between the nitrate of inorganic fertilisers and that from manures, humus, or rainwater. So how can we differentiate inorganic fertiliser nitrates from the organic ones?

Analytical chemists constantly monitor water taken from rivers, lakes, streams, ditches, wells and boreholes, and measure its nitrate content. If the source of water can be linked to a particular stretch of land then it may be possible to say how much nitrate per hectare is being leached. A more meaningful analysis is to measure the water as it actually drains from the soil itself. In Europe and the USA this is done by inserting porous suction cups and sucking out soil water. This is easy and cheap to do, but even then we cannot be certain that water sucked out this way is the same as water which would be leached from the soil. Another way is to take samples of soil and extract all the soluble nitrate, including that which is sticking to the soil particles themselves. But again, we can not know for certain if this method of extraction removes the same nitrate as natural leaching.

A better, but more expensive, way is to cut out blocks of soil and encase them in plastic so that the water leaching from them can be collected. The most expensive method of all is to isolate whole areas with plastic barriers and measure the nitrate run-off while leaving the soil as part of a field. In all these ways we can assess nitrate loss, but none of these methods conclusively proves where the nitrate is coming from.

In order to do this, analytical chemists use a method of labelling the nitrogen atoms with the nitrogen isotope N-15. Nitrogen consists predominantly of isotope N-14, with seven protons and seven neutrons in the nucleus of the atom, giving a mass number of 14. Indeed,

99.63% of nitrogen atoms are of this kind. The remaining 0.37% of nitrogen atoms have eight neutrons in the nucleus, making them slightly heavier with a mass number of 15. If we enrich a nitrogen compound, like ammonium nitrate, with more of the heavier N–15 isotope then we can follow its progress by taking samples and seeing where the N–15 is going. The N–15 isotope is easily detected by an analytical technique known as mass spectrometry.

At Rothamsted they fertilised crops with ammonium nitrate that was enriched with the N–15 isotope and showed that up to 70% of the nitrogen applied as fertiliser was taken up by the crop and harvested in the year it was applied. Of the remainder some was taken up by soil microbes, and some was decomposed back to nitrogen gas by bacteria. Only 5% was leached away into ground water or drained from the land. This loss was higher in the winters following a summer of drought, when the crop did not grow so well and so could not use all the nitrate provided in the fertiliser.

Despite these research findings there is still a call to limit nitrogen fertilisers, maybe because this is the only nitrate input over which we have any real control. There are those who advocate a tax on nitrate fertilisers as a way of discouraging their use, and some have even suggested a tax rate of 50%. Calculations have shown that even with such a tax it would still be worthwhile for a farmer to use nitrate fertiliser, and the price of wheat would need to halve to wipe out the benefits of fertiliser use. A tax might check misuse, especially by those farmers who add more than the recommended amount. A better way to prevent nitrate leaching away would be to make it illegal to apply fertilisers (organic or inorganic) in the autumn, and at the same time to encourage farmers to plant another crop for winter, instead of leaving the land bare and exposed to weathering and run-off.

Removing nitrate from drinking water

Nitrate in drinking water has been described by environmentalists as a 'biological time-bomb' and 'cancer on tap'. Both are alarming phrases, but there is little scientific backing to support them. Nevertheless, they have been the rallying cry of campaigns for the removal of nitrate from public water supplies, even though such nitrate has never killed anyone. Nor is it ever likely to. The World Health Organisation (WHO) has set a limit for nitrate in drinking water of 50 mg per litre (50 ppm), and in 1980 the European Community issued its *Directive on the Quality of Water Intended for Human Consumption* which set the same permitted maximum. This limit is well below the level at which nitrate would pose a risk to health. In those regions where river water is found to contain more than 50 ppm of nitrate, as it does in certain agricultural areas, then it must be treated before it can legally be supplied for human use.

There are two ways of removing nitrate from drinking water. It can be done by biological (organic) methods, or chemical (inorganic) methods. The biological process uses the denitrifying bacteria *Pseudomonas denitrificans* which, in the absence of oxygen from the air, attacks nitrate as a source of oxygen. Having stripped off its oxygen atoms, the bacteria then release the nitrogen to the atmosphere as N_2. Denitrification plants that work in this way are already in operation in parts of Europe where local nitrate levels exceed 50 ppm, and by the end of the century there will be dozens of them in operation. A typical plant treats groundwater, the type that is more likely to exceed the guidelines, and it may process over 6 million litres a day (1.5 million gallons). The denitrifying bacteria need to be supplied with other nutrients, and these can be added in the form of hydrolysed straw[5] and traces of minerals. This requires another treatment of the water after denitrification, in which the organic debris is filtered out. The biological method of nitrate removal appeals to our sense of environmental economy, even though it is costly.

There is a cheaper way of denitrification based on a chemical method called ion exchange, and this is preferred in the USA. As its name suggests, one type of ion, in this case unwanted nitrate (NO_3^-) is replaced with another, such as the safer chloride ions (Cl^-). As the nitrate-rich water filters through the ion–exchange resin the swop takes place in a quick, efficient and easy process. Eventually the ion-exchanger becomes exhausted and has to be recharged with more chloride, and this can be done with a strong

---

[5] The cellulose of straw can be broken down to its glucose components by chemical reaction with water, and this is what the term 'hydrolysed' means.

solution of brine (sodium chloride). This leaves behind a concentrated nitrate solution which has then to be disposed of separately.

Nitrate in the diet

Nitrate in drinking water does not affect its taste, but just knowing it is there worries some people. Those who believe it is a danger to their health think that they can reduce the risk by drinking bottled water. However, not all such waters are free of nitrate. The world's most popular mineral water is Perrier, renowned for the 'purity' of its source, which is at the village of Vergeze near Nîmes in southern France. If you examine the label of a bottle of Perrier you will discover that it contains 17 mg of nitrate per litre. In many areas you would get less nitrate if you drank water from the kitchen tap. Nitrate is to be found in other bottled waters, although generally at much lower levels than this.

Despite the alarms about nitrate we are in no real danger from it, and even if you find a source of water that is nitrate-free, you still cannot avoid nitrate because we produce this chemical naturally in our own body. If you eat a lot of greens, and particularly spinach and lettuce, you may be getting a lot of nitrate.[6] It is not clear why some vegetables take in much more nitrate than they need. Some can have over 2000 parts per million (ppm) of nitrate; for example beetroot can have up to 6000 ppm, and radishes have been found with more than 13 000 ppm, which is 0.13%. These are exceptional examples, and there is no cause to be alarmed by this last figure, which is the highest value recorded in a vegetable, because the amount you eat at any one time will be small. The table opposite shows the more likely amounts in vegetables. The level depends on the soil in which they are grown, and on the time of year: winter-grown vegetables have higher nitrate levels.

Nitrate also comes in foods such as bacon and ham which are cured by being immersed in brines made of sodium or potassium nitrate. The solutions which are used for this have 500 ppm of sodium nitrate or 600 ppm of potassium nitrate, and these figures are low compared

---

[6] It is possible to imagine some people, such as vegetarians, exceeding acceptable limits by eating lots of those foods which have high levels of nitrate.

Nitrate in vegetables

Vegetable	Nitrate/ppm*
Spinach	3000
Lettuce	1600
Red cabbage	600
Carrots	500
Beetroot	500
Celery	400
Green cabbage	200

*Typical values measured in parts per million (ppm), which is the same as milligrams per kilogram.

to the level of 30 years ago when 5000 ppm was the norm. Beer often contains a small amount of nitrate, and because those who drink beer often consume several glasses at one session, they may take in a lot of nitrate.[7]

Sodium and potassium *nitrites* are also used to cure pork and ham, and the use of these is strictly governed by law. In the USA the limits are 120 ppm for sodium *nitrite* and 148 ppm for potassium *nitrite*. *Nitrite* is added to tinned meats, cooked meats, and sausages, not only to preserve them, but also to give them the attractive pink colour that makes them sell better.[8] Adding *nitrite* prevents the bacterium *Clostridium botulinum* from multiplying, which it might otherwise be able to do inside a sealed can of meat or fish. Those eating food from an infected can will get botulism, the deadly paralysing disease that is caused by a toxin which the bacteria produce. *Nitrite* is thought to interfere with the bacteria by cutting off their supply of iron, which is as essential for them as it is for any form of life. *Nitrite* binds chemically to the iron in the meat or fish and prevents it being used by the bacteria.

The amount of nitrate we take in with our food each day depends very much on what we eat. The average person consumes around 75 mg daily, and vegetarians about 200 mg. On top of this there could well be another 30 mg from the things we drink. To make matters worse, our own body produces about 50 mg of nitrate as part of its natural processes, giving a grand daily total intake of around 170 mg

[7] It may even do them some good, since nitrate helps some kinds of kidney stones to dissolve.

[8] *Nitrites* even improve the flavour.

for most people, and 300 mg for vegetarians. Moreover, if we have certain bowel disorders the production of nitrate by bacteria in the gut goes up dramatically.

Should these facts and figures worry us? Thankfully our body can deal with such quantities of nitrate—even 300 mg per day presents no problem. Nitrate passes from our gut to our bloodstream and is quickly filtered out by our kidneys, so that within a day of taking in a large dose we have excreted most of it. We cannot store nitrate within our body so there is no build-up that threatens to be released suddenly, and in any case it is almost impossible to poison someone by giving them nitrate. There is absolutely no need for people to become alarmed about nitrate in drinking water if their supply is within legal limits.

You might still be wondering: why has the WHO set a limit of 50 ppm, if nitrate is not a serious health hazard? And why is this limit being enforced in developed nations? The answer is that *excessive* amounts of nitrate are a health hazard, as we have seen, and tap water contributes to the nitrate which we take in each day. It is really the only source over which we have any control. An upper limit of 50 ppm for water supplies represents a generous safety margin that will protect everyone, even if it is occasionally breached.

Nitrate and human health

People are worried because of what they are told nitrate can do to them or their children. There are two commonly cited ways in which nitrate is said to pose a threat to humans: blue-baby syndrome and stomach cancer.

Blue-baby syndrome

It takes a lot of nitrate to poison a baby. Even if babies were fed water with 100 ppm of nitrate—twice the recommended limit—they would come to no harm. So why has there been concern over nitrate and babies? In fact nitrate has caused the deaths of babies who were fed on milk powder made up with water from wells that had very high levels of nitrate. The circumstances were rather exceptional. The babies turned blue and died because of excess nitrate.

The last case of blue-baby syndrome[9] in Britain was in 1972, and the baby's life was saved. The last fatal case of the disease was in the 1950s. Those at risk were babies in rural areas whose feed was made with water from wells where it was later discovered the level of nitrate could be as high as 500 ppm, ten times the level set by the World Health Organisation. In addition the well-water was contaminated with bacteria which could have exacerbated the problem by causing more nitrate to be produced in the baby's intestines.

If we follow the fate of a nitrate ion when it enters the human body we can understand why babies were at risk. The nitrate itself is harmless, as we have seen, and most nitrate passes easily through. It is the nitrate that undergoes a chemical change which does the damage. Bacteria are to blame, and these remove one of the nitrate's oxygens and so turn NO_3^- into *nitrite*, NO_2^-.

Bacteria in our gut produce about 5 mg of *nitrite* each day. Microbes in our mouths begin the process, which is continued in our stomach and intestines. White blood cells, which engulf and break down invading organisms, also make nitrate and *nitrite*. They use the amino acid **L-arginine** as their source of nitrogen, and they are stimulated to make nitrate by carbohydrate from bacteria. White blood cells are thought to produce nitrate and *nitrite* as intermediates in their production of nitric oxide. This molecule has several uses in the body, from relaxing muscles in our arteries to causing erections in men. It may also be used in the fight against infection. But there can be too much of a good thing, and this is true of *nitrite*.

All animals, including humans, have evolved to cope with a surplus of *nitrite* since it could occur quite naturally. Our first line of defence is vitamin C, which can snatch another oxygen from *nitrite* and so convert it into nitric oxide. We can not destroy all the *nitrite* in this way, and some escapes. It then targets certain of the body's metals, indeed it positively seeks them out and clings to them very strongly. In our bloodstream *nitrite* attaches to the iron atoms at the heart of haemoglobin, the molecule which carries oxygen from the lungs to where it is needed to generate energy. When *nitrite* meets haemoglobin it blocks its active iron site and turns it into methaemoglobin. Still

---

[9] Blue-baby syndrome is known as well-water methaemoglobinaemia in the USA.

we need not worry because we have enzymes to deal with this, and these remove the *nitrite* and so free the haemoglobin.

All these things happen in babies as well, but in relation to their fluid intake, a baby's blood volume is much smaller than that of an adult. An excess of nitrate, and the *nitrite* that is formed from it, will stress their metabolism much more severely. Up to 2% of a normal person's blood may be in the methaemoglobin form. If the level of methaemoglobin exceeds 10% we begin to suffer, and our skin turns a blue-grey. This is what happened with some babies, and hence the name 'blue baby' for those suffering this condition. When the level reaches 40% the baby's life is in danger, and there have been a few rare cases of bottle-fed babies dying.[10]

Breast-feeding is still the best way to nourish and protect a baby, but there is no threat to babies fed on formula foods if the level of nitrate in the water used to make them up is below 50 ppm, the legal limit. The introduction of ready-for-use formula feeds made with nitrate-free water will finally put an end to all worries about blue babies. By the year 2000 these feeds may well have displaced powdered formulas.

Cancer

The other charge levelled at nitrate is that it threatens the health of adults by causing cancer, and especially stomach cancer. The threat comes not from nitrate itself, which is not a cancer-forming chemical, nor from *nitrite* which is also not a carcinogen. The threat comes from nitrosamines, which are a group of known potent carcinogens. These can be produced in the laboratory by reacting *nitrite* with so-called amines, and because nitrosamines are present in some foods it seems likely that the same chemical reaction is producing them there too. **Dimethylnitrosamine** is the most common nitrosamine and is a powerful carcinogen. It is detectable in several foods, although levels vary as the table opposite indicates.

Nitrosamines are extremely reactive and will attack almost any other molecule, including DNA, and this is why they are so dangerous. The table shows what appear to be high levels of these in

[10] Boiling the water to make up a formula feed does not remove the nitrate, but concentrates it.

Nitrosamines in foods

Food	Nitrosamine/ppb*
Hot dog	0–84
Salami	10–80
Bacon	1–60
Smoked fish	4–26
Fried fish	1–9
Cheese	1–4

*Ranges given in parts per billion (ppb), which is the same as millionths of a gram per kilogram.

commonly eaten foods, but the amounts are really tiny, and our body's defences will easily cope with them at the normal daily intake. If you ate a rasher of bacon (50 g) every day for 50 years, and even assuming it had the highest level of 60 parts per billion of nitrosamine, you would still only consume 0.05 g of nitrosamine in all that time.

Nitrosamines are very reactive molecules and the danger is that they will reach the heart of a living cell and attack the DNA. If this happens the damage has to be repaired. This is a relatively simple job that a cell often has to do because of the presence of other more dangerous chemicals called **free radicals**. If the cell's defences are over-whelmed, then the repair job may not be perfect. If this happens we have a potentially cancerous cell. Once this cell starts to replicate itself, and not be recognised as alien, we are in trouble. Luckily we have evolved a whole defence system to protect us against free radicals, and they are quickly rendered harmless by molecules such as vitamins C and E, and these also protect us against nitrosamines.[11]

As we have seen, nitrate is a perfectly normal and natural part of many vegetables, but that does not mean that it could not be a factor in causing cancer. If the sequence: nitrate → *nitrite* → nitrosamines → cancer, really does operate in our gut, then an increase of nitrate in

[11] If you like to start the day with fried bacon and are still worried about nitrosamines, then you can take some evasive action. Microwave your bacon rather than fry it: this produces less nitrosamine because it is quicker and allows less time for them to form. If you fry your bacon, don't dip your bread into the bacon fat because this is where most of the nitrosamines end up. Also drink fresh orange juice (rich in vitamin C), or have some wheat germ (rich in vitamin E).

our diet could lead to an increase of cancers in those parts of the body such as the stomach and intestines.

There are those who seek to link high levels of nitrate in the water supply to the incidence of stomach and colon cancers in particular regions. Attempts to show such a link have met with little success. We might have expected that those most at risk would be the workers making nitrate fertilisers in the chemical industry. This employs a large number of people, and an analysis of their health should indicate whether high exposure to nitrate has led to an increased incidence of cancer. In 1986, the 1300 men who had worked at a fertiliser plant between 1946 and 1981 were studied. Their saliva was sampled and, as we might expect, it showed them to have double the level of nitrate of the rest of the population. Nevertheless they suffered no more cases of cancers of the stomach, oesophagus, bladder or liver than expected. In fact, they were generally healthier than average.

In another investigation, the saliva from patients admitted to hospital with cancer was tested for nitrate, and so was saliva from their families and friends who came to visit them. This ingenious test was designed to measure the nitrate levels in people who were likely to have been exposed to the same levels of nitrate in the drinking water as the cancer patients. The results of this survey revealed that people in areas with high nitrate levels suffered *fewer* cases of stomach cancer than those in areas of low nitrate in drinking water. So much for the nitrate $\rightarrow$ *nitrite* $\rightarrow$ nitrosamines $\rightarrow$ cancer theory, yet the remark that nitrate leads to cancer is still frequently made by those who oppose the use of nitrate fertilisers. We have seen that there may be reasons for opposing the overuse of nitrate fertilisers in farming, but persuading us that this will prevent cancer is not one of them.

Conclusions

Inorganic nitrate fertilisers double or treble crop yields and make farming profitable, and there is no getting away from this fact. If any country were to abandon their use there would be a dramatic fall in crop production and a concomitant rise in food prices. The vulnerable sections of society—the unemployed, the poorly paid, and the old—would be worst affected, since a larger proportion of their income is spent on food.

Longer-term alternatives to nitrate fertilisers are possible. With biotechnology we may even be able one day to design all food crops so they are like legumes, and can fix atmospheric nitrogen. No nitrate fertiliser would then be needed. The prospects of doing this are still very remote.

Nitrate is a perfectly natural part of the world around us. We take it into our bodies, we make it there, and we dispose of it easily. Even that which turns into *nitrite* appears to pose no real threat, although this is the precursor of nitrosamines in some foods. Nor are babies at risk from nitrates, even if they are bottle-fed, as long as the water does not exceed the WHO limit of 50 ppm, and it rarely does.

You may still want to take some action as an individual that might help reduce nitrate levels in rivers, and thereby prevent their being choked with water weeds and algae. I can give you no easy advice. The best course of action might seem to be to stop eating meat. Growing crops to feed farm animals for slaughter is a very inefficient process. Less nitrate would be used if humans took in more of their protein as vegetable protein rather than animal protein. Yet to grow more crops we need to plough up more grassland, and as we have seen that results in even more nitrate draining into rivers. And if you become a vegetarian you take in more nitrate anyway. Happily, this sacrifice would appear to pose no serious threat to your health.

CHAPTER NINE

Carbon Dioxide

Carbon dioxide, CO_2; greenhouse gases; the global CO_2 budget; global warming; the good side of carbon dioxide

IF you ask someone which chemical presents the greatest threat to the environment, they would probably reply 'carbon dioxide'. Which would not surprise us, because we are continually warned that this is the greenhouse gas which will cause global warming in the next century, and in so doing will make parts of the Earth uninhabitable. Most people are probably not aware that such warnings and predictions are based solely on computer simulations, nor are they told that there has already been an increase of over 25% in the level of atmospheric carbon dioxide this century, with no reliable signs of global warming. Although some environmentalists find this rather puzzling, we can explain the anomaly with a little chemical understanding.

Carbon dioxide is produced every time we burn something containing carbon, whether it be a cigarette, a bonfire in the back garden, gas in our heating system, or petrol in our car. If we switch on a light we produce it because we consume electricity, and this is generally produced by burning coal, oil or gas in a power station. Almost everything we do produces more CO_2.

Carbon dioxide is a greenhouse gas, of that there is no doubt. But if we take a closer look at the gases of the atmosphere, including carbon dioxide, we discover that CO_2 has already contributed almost all it can to global warming and is really a spent force. Its ability to cause further warming is debatable. This chapter will take a closer look at carbon dioxide itself, and explain why it, along with other molecules, is called a greenhouse gas. We will consider the various issues which surround this controversial chemical. How much of this gas is there in the atmosphere? How much are we adding each year? Where does it all go?

We will also look at other less well-known aspects of this chemical, such as its role as an industrial resource.

Carbon dioxide, CO_2

There is a lot of carbon dioxide in the Earth's atmosphere—about 2750 billion tons in total—but the concentration in air is quite low, about 0.035% or, as it is more usually reported for amounts this small, 350 parts per million (ppm) by volume. Most CO_2 in the atmosphere is released from natural sources such as the soil, the sea and subterranean depths. Indeed, the human contribution to atmospheric CO_2 is small by comparison. At the beginning of the Industrial Revolution 250 years ago the concentration was around 280 ppm. This amount did not change significantly until about a century ago when we began to exploit oil. It started going up more steeply about 40 years ago with rapid industrialisation around the world and the growing popularity of the motor car, of which there are now over 400 million. Atmospheric CO_2 levels are rising by about 1 ppm per year, but only a fraction of the CO_2 we emit to the atmosphere each year remains there: some is taken up by plants, some is absorbed by the sea, and a lot dissolves in rainwater. It becomes part of the carbon cycle of Nature.

The amount of CO_2 extracted from the atmosphere by living plants is estimated to be about 350 billion tonnes per year, with a roughly equal amount passing back as these are eaten by animals, or as they die and decay. A similar amount of around 350 billion tonnes of CO_2 probably moves back and forth between the atmosphere and the sea each year. It is against these massive changes that the human release of 22 billion tonnes a year from fossil fuels seems relatively unimportant, but over a century or so this could become significant and might make all the difference in tipping the global environment into an unstable state. Later in this chapter we will look at the various compartments of the environment and the amount of carbon each contains.

Carbon dioxide is not in itself a dangerous chemical. Indeed, as a refrigerant it has probably saved many lives, and those who have died from CO_2 have been killed by natural CO_2 erupting from deep below the surface of the Earth. It was from a subterranean source, below Lake Nyos in western Cameroon, that massive bubbles of CO_2 suddenly surfaced one night in August 1986. Hundreds of sleeping villagers living near the lake, and thousands of their animals, died from asphyxiation as tonnes of the gas emerged from the lake and rolled over the surrounding countryside.

Less threatening is the underground 'lake' of CO_2 a mile below the surface near the extinct volcano Mount Gambier, in South Australia, which is being commercially exploited. In the USA many wells of CO_2 have been discovered while drilling for oil, and the gas from some of these wells has been put to use. It is piped hundreds of miles to oil fields where it is re-injected into oil-bearing rocks to force out more oil from wells that were thought to be exhausted. The Western Texas oil-field will eventually yield an extra 56 billion barrels of oil this way.

You might wonder why there should be large deposits of CO_2 underground. Part of the answer is a natural release of gases from the molten rock deep below the crust, and which we observe when a volcano erupts. However, a lot of CO_2 in the Earth's crust is of biological origin, and comes from prehistoric plants. Their organic matter that gave rise to coal, oil and natural gas was mainly carbohydrate, which consists of carbon, hydrogen and oxygen. It underwent the chemical transformation due to the heat and pressure in the Earth's crust, when some of the carbohydrate ended up as hydrocarbons, and some as CO_2.

Billions of tons of carbon are cycling through all parts of the environment, and have been doing so for hundreds of millions of years. In our pre-industrial state the human race was a part of the natural order, but then we started tapping the energy of fossil carbon that was locked away in the Earth's crust. Since the Industrial Revolution began in the mid-18th century, we have unlocked vast reserves of fossil carbon, almost always to burn it. We have done this on such a large scale that we have increased the amount of CO_2 faster than other natural processes in the cycle could dispose of it.

Early analytical chemists who measured the concentration of CO_2 in the air found it varied considerably with the time of year and the location. The limits of accuracy of their analyses were such that they could not register small global changes. An agreed reference point was needed and this is now the value published regularly by the monitoring facility at Mauna Loa in Hawaii.[1] This started reporting CO_2 levels in the 1950s, and for almost 40 years scientists there have observed the seasonal ups and downs in the level of CO_2 in the atmosphere of the mid-Pacific. These have shown an overall upward trend, rising from 315 ppm in 1958 to over 350 ppm today. Environmentalists blame this increase in CO_2 on two main factors: industrialised nations using fossil fuels, and Third World countries burning forests to clear land for agriculture. As the change seemed always to be upwards, people quite naturally began to wonder what effect it might be having.

Greenhouse gases

Of the energy from the Sun which falls on the Earth, less than half warms the surface. The clouds and oceans reflect about 30% straight back into space. Another 25% is absorbed by the atmosphere itself. The remaining 45% which heats the planet also has eventually to be reflected back into space as radiant heat (more correctly called infrared rays), otherwise as more and more heat is trapped the planet would get hotter and hotter. The radiated heat, however, has to pass back through the atmosphere, and molecules in the air are able to trap it and reflect some of it back down again. This is the greenhouse effect, and the molecules which behave this way are called greenhouse gases, of which CO_2 is the best known. But CO_2 is not the most important—water vapour is.

The effect of trapping the Sun's energy for heat is one that we are familiar with in a normal greenhouse. The glass allows rays from the Sun to enter, but prevents the warm air from escaping. An analogy

[1] Some scientists have questioned the wisdom of siting such an important global facility at Mauna Loa, because it is located on the side of a volcano which erupts on average every three and a half years, and it is only 27 km from the largest active volcanic crater in the world.

can be made with the planet; if molecules in the atmosphere stop radiant heat escaping, this too will warm up. It was a little misleading of the proponents of global warming to speak as if releasing CO_2 was the equivalent of building a greenhouse round the planet, although this was the image often used to drive home their message. In fact we *need* the atmosphere to act as a greenhouse, to prevent too much heat escaping into space, otherwise we would experience daily fluctuations of temperature like those of the Moon, and that would make life impossible.

The atmosphere consists of several gases, of which **nitrogen**[2] and **oxygen** account for 99% of the total volume, and **argon** accounts for most of the remaining 1%. The other gases are there in only tiny amounts, and for this reason we do not talk of them in percentage terms but in units of parts per million (ppm). Of the lesser gases, the most abundant appears to be CO_2 at 350 ppm, and this is what you will find if you consult tables which give the composition of the atmosphere. You will not find water vapour mentioned because such tables will be referring to samples of *dry* air. Water vapour is excluded because the amount varies so markedly, and it ranges from a fraction of 1% on a very cold day, to 4% on a hot, humid day. The table opposite gives two sets of data for the gases of the atmosphere: the 'dry air' one which is commonly quoted, and a 'normal air' one in which I have recalculated the concentrations assuming there is 2% water vapour present. You will see that water vapour is really the third most abundant gas in the atmosphere. It is also the most powerful greenhouse gas.

The Earth's atmosphere keeps the temperature reasonably stable, smoothing out the daily changes, and the seasonal ones. The vapour which does this so effectively is water. Without water, the average temperature of the planet would be about 33 °C colder than it is. Water is very good at trapping a lot of the infrared rays which would otherwise be lost to space, and water vapour alone is responsible for a warming of roughly 32 °C, or about 97% of the greenhouse effect. Carbon dioxide helps a little bit as well, accounting for about 1 °C.

---

[2] A reminder that when the name of a substance appears in **bold** type, it means that there is more about its chemistry in the Appendix.

The composition of the atmosphere (parts per million by volume)*

	dry air	normal air**
nitrogen	781 000	766 000
oxygen	209 500	205 000
water	0	approx. 20 000
argon	9500	9000
carbon dioxide***	350	340
neon	18.2	17
helium	5.2	5
methane	1.7	1
krypton	1.1	1
hydrogen	0.6	0.5
nitrous oxide	0.3	0.3
carbon monoxide***	0.2	0.2
xenon	0.1	0.1
ozone***	0.03	0.03
CFCs	0.000 76	0.000 76

*The figures in these columns do not total to exactly 1 000 000 because of variations in the values of some of the gases and limitations to the number of significant figures for the more abundant gases. **For normal air I have assumed a figure of 20 000 for water vapour and adjusted the figures for nitrogen, oxygen, argon, etc. accordingly. ***There are seasonal variations in the amounts of these gases.

What makes a molecule act as a greenhouse gas? The answer to this question is one of chemistry: a molecule can absorb radiant heat (infrared rays) only if it has a certain kind of structure. To be a greenhouse gas, a molecule needs to have chemical bonds, because these vibrate with the same frequencies as infrared rays and can absorb them. A gas like argon (9000 ppm) consists of just single atoms and has no chemical bonds, so it cannot interact with infrared rays. The same is true of the other gases which consist of single atoms: **neon** (17 ppm), **helium** (5 ppm), **krypton** (1 ppm) and **xenon** (0.1 ppm).

Even if a molecule has a chemical bond it does not necessarily mean that it will absorb infrared rays. Those gases in which there is a single bond between two atoms of the same element can not interact with radiant heat, and this is why nitrogen N_2 (766 000 ppm), oxygen O_2 (205 000 ppm), and hydrogen H_2 (0.5 ppm) are not greenhouse gases either.[3] They have chemical bonds, but not the right kind.

---

[3] When a molecule like these vibrates there is no change in the electric field surrounding it. Without such a change it can not absorb infrared rays.

To absorb infrared rays, a molecule of two atoms must have both atoms different, such as carbon monoxide, or it must have three atoms or more. The ones in the Earth's atmosphere which therefore qualify as the chief absorbers of heat are: water (H_2O), **carbon dioxide** (CO_2), **methane** (CH_4), **nitrous oxide** (N_2O), **carbon monoxide** (CO) and **ozone** (O_3). There are lots of other molecules which qualify, such as **sulfur dioxide**, **methyl sulfides**, and the **CFCs**, but these are present only in tiny amounts.[4]

The wave band of the infrared range spans the wavelengths from 1 micron to about 40 microns (a micron is a millionth of a metre). How well a greenhouse gas reflects back radiant heat depends on how much of this range it covers. Different vibrations of a molecule occur at different wavelengths, and for a particular molecule we get a particular pattern of bands. These bands are determined by the chemical structure and are unique to each molecule. Water has many bands at many wavelengths. CO_2, on the other hand, has only two.

The infrared absorbance of the atmosphere can be recorded by passing a beam of infrared light through a sample of air and noting the wavelengths that are being absorbed by the greenhouse gas molecules. When we examine the spectrum we discover that at many wavelengths the energy is being absorbed mostly by water, whose bands blot our whole ranges of the spectrum. CO_2 absorbs radiant heat very well, but only at 4.3 microns and 15.0 microns. It traps almost all the heat being emitted at these wavelengths, but only at these wavelengths—which is why it contributes relatively little to global warming. There is a 'window' in the spectrum through which radiant heat can escape, and this is the 8–13 micron range. None of the natural greenhouse gases absorbs at these wavelengths.[5]

Once we understand that the absorbing capabilities of CO_2 are rather restricted by its chemistry, we should not be surprised that

---

[4] Even the HFCs make excellent greenhouse gases, and these are the hydrofluoro-carbons that are being introduced to replace the CFCs which damage the ozone layer. The concentrations of HFCs are too small to be included in the table on p. 239.

[5] On the other hand the CFCs do absorb in this region. For example CFC-12 absorbs mainly in this region, as do most of the other molecules in which there are carbon-to-chlorine chemical bonds, such as the organochlorine solvents.

there is little relationship between the amount of this gas in the atmosphere and the average global temperature. This is why the 25% increase in CO_2 in the atmosphere during the past century can only have contributed a fraction of a degree to raising the average temperature of the Earth, and why this has not been detectable so far.

It would take about 200 years at the present rate at which CO_2 is increasing for the level of CO_2 to double to around 700 ppm. This will have an effect, but at most it will add another 1 °C to global warming. There is no way could it boost planetary temperatures by the 5 °C that some have predicted will happen in the next century. There is no way that CO_2 can behave like water as a greenhouse gas and absorb heat at many wavelengths—its chemistry will not allow it to do this.

So why was there so much alarm about CO_2? This arose through some data which seemed to show a strong link between this gas and previous global temperatures. Analyses of ice cores from deep within the polar ice-cap revealed that during the ice ages the amount of CO_2 fell, and was lowest during the coldest periods. Between the ice ages it rose, and so did the Earth's temperature. Supporters of global warming saw this as proof that the amount of CO_2 in the atmosphere was the main regulatory mechanism. The ice-core data provided what they badly needed: experimental evidence to support their computer simulations. Then in 1992 it was reported by a group of Norwegian and Japanese scientists that these ice-core data were flawed. The crude way in which the ice samples were taken meant that the chemical analysis of the CO_2 trapped in the ice was unreliable. More careful collecting of samples and more accurate analyses showed that the CO_2 trapped in the ice changes over time, and varies with the condition of the ice itself.

In the summer of 1992 more experimental results, reported in the journal *Nature*, further undermined the supposed relationship between CO_2 and global warming in the past. A new ice core had been drilled over 3000 metres down into the Greenland ice sheet, and its water had been analysed. This study had been carried out by a group of researchers from all over Europe under the auspices of GRIP, the Greenland Ice Core Project, and they had sampled snow that had fallen over the past 260 000 years, a period which included two Ice Ages and three interglacial warm periods, including the present one. The current warm period is called the Holocene era, and so far has lasted about 11 000 years. The previous ice age lasted about 100 000 years. This was preceded by the Eemian warm period of 20 000 years, and before that another Ice Age of 90 000 years.

The ice samples corresponding to the Eemian warm period were particularly revealing. During the hottest years of the Eemian period, the Earth was about 4 °C warmer than it is now, and there were lions and elephants roaming Europe. During that era temperatures sometimes fell dramatically by as much as 10 °C over a period as short as 10 years, and then the Earth remained cold for several decades or centuries before suddenly warming again. Global warming came and went several times in the Eemian period, until finally there was a warm spell of 2000 years. Such rapid changes can not be linked to a doubling, a tripling, nor even a quadrupling of the amount of CO_2 in the atmosphere. But whatever the cause, it did not last. The Earth plunged into the last Ice Age, which it did with a rapid drop in temperature of about 14 °C in only ten years.

You may be wondering how the water molecules of compacted snow which fell a hundred thousand years ago could preserve a record of the Earth's temperature. This information is locked away in their oxygen atoms. The GRIP core was studied by measuring the isotopes of oxygen, a method developed by Willi Dansgaard of the Niels Bohr Institute at the University of Copenhagen. Oxygen consists of two isotopes, oxygen-16, which accounts for 99.8% of all oxygen, and oxygen-18, of which there is 0.2%. Oxygen-18, as its name implies, has a mass of 18 because it has two extra neutrons. Slightly more energy is needed to evaporate water with oxygen-18, and the result is that as the climate gets colder, the amount of this in snow decreases slightly. Chemists can measure the oxygen-18 to oxygen-16 ratio very accurately using a mass spectrometer, and relate this to the temperature of the winter in which the snow fell; the colder the winter, the lower this ratio.

Our present interglacial warm period, the Holocene, has not been without its ups and downs. A Little Ice Age from the sixteenth to the nineteenth centuries saw the Alpine glaciers advance and winter fairs held on a frozen Thames in London and ice-bound canals of Holland. Indeed, the threat of a new Ice Age was the concern of many environmentalists and meteorologists in the 1970s, who based their predictions on falling global temperatures recorded from 1940 to 1970. This happened despite the rising level of CO_2 in the atmosphere.

Nevertheless, in the 1980s the Earth appeared to be warming and the earlier ice-core CO_2 data seemed to be proof that this warming could cause a global climate change. In the 1990s it appears to be cooling, and there were several unusual happenings: Hudson's Bay stayed ice-bound throughout the cold summer of 1992. There were snowstorms in Florida in early 1993, followed by another cool, wet summer, and record low temperatures in the winter of 1993/4. Of course this kind of local evidence is as unrepresentative as the heat-waves and droughts of the 1980s. What is required is a *global* average.

It is easy to measure accurately the temperature of the Earth at any one point, but it is not easy to turn millions of such recordings into an average for the whole Earth, most of which is covered by sea. A satellite, on the other hand, can take a temperature picture of the whole planet, and much store was set on the data this would provide. In the summer of 1993 Dr John Christy of America's space agency, NASA, reported a 14-year study. Despite extensive searches using the most advanced technology, their satellite had found no global warming between 1979 and 1993. Apparently the rising level of CO_2 was still having no effect.

This is reassuring, but not unexpected from what we know of CO_2. However, we should not become complacent because there are other gases being added to the atmosphere that can absorb radiant heat at wavelengths in the 8–13 micron window which currently allows heat to escape from the Earth. One such gas is methane, which comes mainly from natural sources, but is also building up in the atmosphere as a result of human activity. Some of this gas escapes from leaks in gas pipelines, some is generated by rotting organic matter in landfill sites, some is given off by paddy fields, termite mounds, and some is produced in the stomachs of cows and in the intestines of many domestic animals. Indeed, humans also add to the methane in the atmosphere in this way.

Other volatile hydrocarbons will also absorb infrared heat in the same region of the spectrum, as will any gas with carbon–hydrogen bonds. When you fill up your car at a petrol station, the incoming petrol will displace the same volume of hydrocarbon-saturated air from the fuel tank, and this will then act as a greenhouse gas. Petrol companies are already alert to this danger and are now finding ways of reclaiming these lost hydrocarbons by installing air filtering units at fuel depots and even at service stations.[6]

Hydrocarbons cannot linger for long in the atmosphere because they are susceptible to reaction with the oxygen of the air. Other greenhouse gases that we are producing will linger longer, such as the CFCs. The countries who signed the Montreal Protocol of 1987 agreed to the phasing out of the CFCs by the end of the century, and they agreed to speed this up at their London meeting in 1990. These gases are being controlled not because of their contribution to global warming, but because the chlorine they contain threatens the ozone layer. The gases which are set to replace them, the **hydro-fluorocarbons** (HFCs), do not contain chlorine so at least they are not able to deplete the ozone. They could still contribute a little to global warming though.

The global CO_2 budget

Even though CO_2 may not be the greenhouse threat which we were led to believe, it is still pouring into the atmosphere. In some respects this is a measure of human folly. There are now far too many of us, and as we go about our business, building, farming, burning, and wasting, we are producing CO_2. Our great grandchildren may not blame us for global warming by doing this, but they might blame us for a global squandering of resources that has achieved little of which we can be proud.

All the carbon that is locked away in deposits of gas, oil and coal was once CO_2 in the Earth's atmosphere. We saw in the last chapter that there is a nitrogen cycle in Nature, and similarly there is a carbon cycle which circulates this element through the environment, and through all living things. We can trace the route of a carbon atom as it moves through this cycle. It starts with CO_2 in the atmosphere, and then with the help of sunlight and the chlorophyll in the leaves of plants it reacts with water to form carbohydrate. This may then move

---

[6] When they do they can trap the petrol that evaporates and sell it again.

through the cycle within a few weeks by being eaten and used as a source of animal energy, and as a result the carbon will be breathed out again as CO_2. On the other hand it may be incorporated into other molecules and remain locked away for a year, a century, or even a million years or more, before it returns to the atmosphere.

When humans evolved they had little impact on the environment until they started felling and burning trees, draining land and farming, but even then they only speeded up the cycle without increasing the amount of carbon moving through it. Humans started to make a significant contribution when we began burning fossil fuels and greatly increasing our numbers. What effect are we having?

Human activity generates CO_2 in six important ways. Three of these return the CO_2 whence it came, to the air, and the other three add CO_2 to the atmosphere from sources that were locked away underground. The first three are breathing, rearing farm animals and using wood. The second three are making cement, burning fossil fuels and draining land for cultivation. Together these add up to a yearly total of over three million litres of CO_2 per head of population. This is about the capacity of a large hall of $30 \times 10 \times 10$ metres (roughly $100 \times 30 \times 30$ feet). If you prefer to think in weights of CO_2 rather than volumes then this is six tonnes per person (see the Box). In some regions the release per person is less than a million litres per year, while in others, such as the USA and Europe, it is over ten million litres.

How to measure CO_2: units of volume or weight?

Gases, like liquids, are measured in volumes, and the litre is a convenient measure of volume that most people can understand since it is the size of a mineral water bottle. Americans may not be as familiar as Europeans with this quantity, but a rough equivalent is a quart. This is about 5% smaller than a litre. When volumes get larger it is more convenient to talk in terms of a cubic metre, which equals 1000 litres. The cubic metre is 1.307 cubic yards, and this volume is the same as 880 quarts. With even larger volumes the cubic metre measure is not adequate, for example when we are considering global totals of CO_2. Then it is easier to talk in terms of weight, in this case in metric tonnes. A tonne of CO_2 occupies 500 cubic metres or 500 000 litres. A tonne is about 10% larger than a US ton.

Let us begin by looking at the CO_2 which we produce in our own bodies. Most of the food we eat provides us with energy, and this comes mainly from carbohydrates and fat. The average person needs 2200 kilocalories of energy per day from food, and if this were all derived from carbohydrate then we would need to consume 550 grams (20 ounces). Our yearly requirement would be 200 kg. Of course we cannot live only on carbohydrate, but for the sake of our calculations let us assume that all our energy is coming from this source.[7]

The molecular weight of a carbohydrate, formula $C_6H_{12}O_6$, is 180 units, and when this is burnt or used as fuel in our bodies it ends up as six molecules of CO_2 which together weigh 264 units. Therefore if we consume 200 kg of carbohydrate a year this would end up as 300 kg of CO_2 expelled from our lungs. In volume terms this is 150 000 litres, which works out about at 17 litres per hour. This tallies with the observed rate of exhalation.

We could do the same sort of calculation for domestic animals, who owe their very existence to human needs. The number of these of course probably exceeds the number of humans by quite a margin, so it is not easy to estimate how much CO_2 they are breathing out each year. One reasonable estimate is that it amounts to an annual total of 2 billion tonnes, which is 370 kg, or 185 000 litres, for every human being on the planet.[8]

But the human activity which returns most CO_2 to the atmosphere is our use of wood. Again it is difficult to put a figure on the amount of wood that is cut down each year, and estimates vary between 3.5 billion metric tons and 7 billion metric tons. Most of this is harvested in the Third World where it is used for cooking, which of course returns it immediately to CO_2. Even the wood that is used in industrial countries as paper products and building material eventually ends up this way—fairly quickly if our waste goes to a municipal incinerator, more slowly if we recycle it, and very slowly if it ends up in

[7] We could do a much more sophisticated calculation based on our total diet, and we could actually measure the amount of CO_2 we evolved during the course of a whole day, but as we are only talking in round figures for the average human, precise measurements on an individual are not much help.

[8] These figures are not as exact as they appear, and are based on the conversion factors in the Box on p. 245. Another way of expressing them would be to the nearest round number, i.e. 400 kg or 200 000 litres.

landfill sites and decomposes over the centuries. We may delay its return by using it for books, furniture or house-building, but even these are finally burnt or rot away. If we assume the destruction of 5 billion metric tons of wood per year, this will turn to 7 billion tons of CO_2 when burnt, or 1.3 tons per person, which in volume terms is 650 000 litres.[9]

The CO_2 from our food, our animals and our use of wood came from the air in the first place, so all we are doing is putting it back. If humans didn't do that other creatures might, so we are not upsetting the carbon cycle by these actions: this release of CO_2 is merely recycling it. The 'natural' CO_2 adds up to about a million litres per person per year, and I have given these figures so that we can put into perspective the 'new' CO_2 that we add to the atmosphere each year.

Let us now turn to the CO_2 we are releasing from carbon in the Earth's crust, the CO_2 that has been long locked away. There are large deposits of mineral carbonates such as limestone (which is **calcium carbonate**) and **dolomite** (which is mixed magnesium and calcium carbonate). Cement works roast a billion (1000 million) metric tons of limestone a year and this drives off 440 million metric tons of CO_2. A further 200 million metric tons of limestone is heated in kilns to make lime for agriculture and industry, and this adds another 90 million metric tons, making a total of 530 million metric tons of CO_2 in all. Consequently cement and lime manufacture is adding about 50 000 litres per person per year to the atmosphere, and this does not include the fuel burnt to do the roasting. That is included in our next calculation.

The amount of CO_2 from fossil resources is easy to compute since a careful tally is made of the natural gas, oil and coal that is used each year. For example the *BP Statistical Review of World Energy for 1993* reveals that together these totalled 7073 million metric tons of 'oil equivalent'. This is made up of 3128 million tons of oil, 1781 million tons of natural gas, and 2164 million tons of coal. Natural gas is mainly methane, chemical formula CH_4, oil is a complex mixture of hydrocarbons of average chemical composition CH_2, and coal is even more complex, has even less hydrogen, and is roughly CH. To have every-

[9] The calculation is the same as for carbohydrate, which is what wood mainly is.

thing reduced to 'oil equivalent' means our calculation of how much CO_2 is given off is easy. When CH_2 turns to CO_2, then 14 tonnes of 'oil equivalent' turns into 44 tonnes of CO_2.

The release of CO_2 from fossil reserves in a developed country comes in roughly equal proportions from power stations, transport, industry and heating. Of course, not all fossil fuel is burnt, but by far the major part is, and even that which is turned into plastics, textiles, and paints eventually suffers the same fate as the products from wood, and ends up as gas in the atmosphere. The average fossil fuel used per person is 1.75 metric tonnes per year, which converts to 4.1 tonnes of CO_2 or 2 050 000 litres—twice the amount we release from 'natural' sources. The amount of CO_2 per head of population from fossil fuel varies considerably from country to country. The countries with the highest totals are the US and Canada, where it is about 15 tonnes per person per year, and Europe and Japan, where it is in excess of 7 tonnes.

Finally there is an unknown amount of CO_2 we release by our draining of land for agricultural use. Wetland locks up a lot of organic matter as humus, which slowly turns to peat and eventually would end up as other fossil fuels if it were to experience pressure and temperature as a result of geological action over millions of years. As far as I am aware no one has yet put a figure on the CO_2 from the land itself, but intensive agriculture slowly reduces the organic content of soils, and this is mainly carbon based. The agricultural release of CO_2 from this source might indeed be quite large, but we can not include it in our calculations—all we can do is remember that what follows is an underestimate of the amounts of CO_2 released each year as a result of human activity.

If we total all the above sources we find the average human is responsible for not less than 3 million litres of CO_2 being emitted into the atmosphere each year. The inventory is as follows:

	litres
Breathing	150 000
Rearing farm animals	185 000
Using wood	650 000
Making cement	50 000
Using fossil fuels	2 050 000
Total per person	**3 085 000**

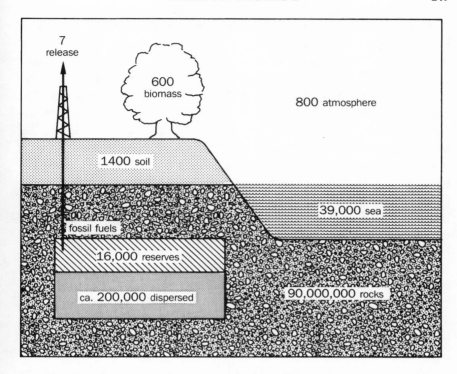

The Earth's carbon. The amounts shown are in billions of tonnes of carbon (10^9). The carbon in the atmosphere is mainly CO_2, that in the sea and rocks is carbonate, and the fossil fuel forms are mainly hydrocarbons or coal.

This works out at 6.2 metric tonnes per person, giving a grand total of 33 billion tonnes per year for the planet as a whole. At this point you may be getting a little weary of large numbers and you still do not know how this number compares with the overall distribution of carbon on the Earth. The diagram above shows the amounts of carbon in the major compartments of the environment. This information is expressed as weights of carbon to take into account all the varied forms in which carbon exists. For the atmosphere this means talking of carbon even though it is there as CO_2.[10]

[10] It is easy to convert from one to the other because a tonne of CO_2 contains 0.27 tonnes of carbon. The 2750 billion tonnes of CO_2 in the atmosphere represent 800 billion tonnes of carbon.

The amount released to the atmosphere from fossil fuels is 7 billion tonnes per year, adding to the 800 billion tonnes already there in the atmosphere. This is roughly 1%, and yet the amount in the atmosphere goes up only by less than a third of this each year. The reason is that CO_2 dissolves in the sea, and is washed out in rainwater. Most importantly of all, the plants of the planet, the so-called biomass, convert CO_2 to the chemicals that are needed by living cells. This also takes place in the upper layers of the sea and it is now thought that the amount of carbon dioxide these absorb has been greatly underestimated. The transfer of carbon to other compartments removes most of the extra carbon added from fossil fuels, but it is not clear where a lot of CO_2 is ending up. Some obviously could be going into increased growth by plants and especially the trees of the forests which still cover large tracts of the Earth. A fully mature forest does not have a net input of carbon dioxide because there is a balance between growth, which absorbs CO_2, and respiration and decay which release it. But if the level of CO_2 in the atmosphere increases, so does plant growth. This is one positive effect of CO_2 which is made use of in commercial greenhouses, as we shall see.

The level of CO_2 in the atmosphere is rising, and we can link the increase to the burning of fossil fuels. This can go on at the present rate for a long time yet, but reserves are finite, even though the success in finding new ones more than keeps pace with the rate at which they are being used up.[11] One day they will become too expensive to extract, however advanced our technology, but that might be a thousand years in the future. Then we may stand condemned for the sheer waste of so much of it. Future generations may well ask why we used these fossil reserves to such unworthy ends as heating buildings that were badly insulated, generating electricity in such an inefficient manner, or driving a car when we could walk or cycle.

---

[11] If the 1000 trillion tonnes of oxygen in the Earth's atmosphere all originated from photosynthesis by plants of CO_2 then we can calculate that the amount of carbon locked away in the crust could be an astounding 375 *trillion* tonnes. Much of it is too dispersed to use as a fuel.

Global warming

An Atmospheric Temperature Workshop held in the USA in 1991 issued a report which contained the following warning: 'The specific concern today is that the exponentially increasing concentrations of anthropogenically introduced greenhouse gases will, sooner or later, irreversibly alter the climate of the Earth and thereby disrupt global weather distribution, agricultural production, water supplies and other economic and social activities.' Put in plain language, this says that because humans are releasing greenhouse gases in ever-increasing amounts, sooner or later we will damage the Earth's climate, causing droughts, floods, famine and economic ruin. The authors made these doleful forecasts on the basis of calculations carried out on 'the most powerful supercomputers available.' The meeting concluded that the most pressing need was to run even bigger computer simulations, and to discuss the data from these at more workshop meetings around the world.

When computer modellers took it upon themselves in the late 1980s to predict what might happen if the level of CO_2 in the atmosphere were to double, they came up with a forecast of global warming of 5.2 °C. If the level of CO_2 were to rise by 1% per annum this might occur within 70 years, they said. Such a rapid change in climate, over so short a period, would melt polar ice and flood the low-lying areas of the globe.[12] The world was warned of the catastrophes that lay ahead. Only a few years earlier there had been alarms about global cooling and the imminent ice age. This was rather unfairly recalled by Douglas Story in a letter to the American Chemical Society's weekly magazine *Chemistry & Engineering News* in September 1993:

'I am concerned about our public integrity when I see the people who lectured in the early 1970s at my university about the coming environmental disaster of global cooling, now lecturing (and making a good living at it) about the coming disasters of global warming. It scares me to see other scientists and the media starting efforts ... to substantiate the theories of these guys. It petrifies me to know that government bureaucrats and politicians are developing social and business policies based upon this camera-lens science.'

[12] Only ice sheets on land can raise sea levels when they melt. If all the ice at the North Pole, which is sea-ice, were to melt, it would have no effect on sea levels. Ice floats on water because it is less dense than water and when the floating ice melts, the level of liquid does not change.

Those earlier warnings of an approaching ice age were overtaken by events when the global temperature, which had been falling from 1940 to 1970, suddenly went up a little. The message of doom changed accordingly: the world was now in danger, not of freezing in the next century, but of overheating. The prophets even suggested why this might happen: there was too much carbon dioxide going into the atmosphere.

The present concerns about global warming really began in the summer of 1988 when predictions based on computer-simulations were published. These forecast that the world was about to warm up rapidly. There would be worldwide floods as the ice at the poles melted and sea levels would rise by as much as 8 metres (25 feet). This message of a Second Flood was taken up by the media, and soon even politicians were voicing alarms of global warming. Paperback books appeared, explaining the message in graphic detail. Some environmentalists and meteorologists warned of mass evacuations of people as Pacific islands disappeared, and mass migrations of refugees from Bangladesh and other low-lying regions of Asia, which are among the world's most populated regions.

The main protagonist behind all this agitation was The Intergovernmental Panel on Climate Change (IPCC). In 1988, members of the World Meteorological Organization and the United Nations Environment Programme set up the IPCC to assess the available scientific information on climate change, and the environmental impact this might have. The IPCC formed itself into three committees, one to look at the evidence for climate change, a second to forecast the results of such changes, and a third to suggest what could be done.

The first of the committees was the most important because its findings would determine the validity of the others. The so-called Working Group I published its deliberations in July 1990. The report was the work of 170 scientists in 25 countries.[13] The IPCC blamed CO_2 for being the main contributor to the enhanced greenhouse

----

[13] The report admitted that there were some scientists who did not agree with the IPCC findings. Their views were ignored. However, a survey of climatologists in 1991 showed that fewer than 20% of them believed in imminent global warming.

effect, along with other gases such as methane, nitrous oxide and the CFCs, and concluded that an immediate 60% reduction in the emissions of these gases was necessary to prevent them from further increasing in the atmosphere. The IPCC predicted that if we continued to emit these gases at the present levels, then the temperature of the Earth would rise by 3 °C in the next century, and this would lead to a rise in sea level of 65 cm. The report admitted there were uncertainties in their predictions because their computer simulations could not take into account clouds, oceans and polar ice sheets.

The doubts and uncertainties were ignored by environmental pressure groups and their supporters, who promoted the message of the IPCC with gusto. No publicity was given to the warning in the IPCC report that it was 'neither an academic review, nor a plan for a new research programme.' The environmentalists behaved as if the report was an official document issued jointly by the world's governments, as its very name seemed to imply. Coming on top of a series of warm winters and hot summers in the Northern Hemisphere, the IPCC report seemed to confirm what people were really experiencing: the world was rapidly warming up. Regrettably the scientists of the IPCC did nothing to check this alarmist scenario; indeed, by not refuting their wilder supporters they appeared to agree with them. Some popular science writers gave their imaginations full throttle, and talked of runaway global warming with the world eventually heating to almost 500 °C (the 'Venus scenario')[14], while others spoke of the Western Antarctic ice sheet suddenly sliding into the sea causing an instant tidal wave that would sweep the world and cause the sea level to rise by 9 metres (30 feet).

The 1990 IPCC report calculated that as more CO_2 from fossil fuels forced up the temperature, water vapour would become a threat in that it would provide positive feedback, thereby enhancing global warming. As the temperature went up, more water would evaporate from the oceans. This would increase the greenhouse effect, which in turn would encourage more water to evaporate, and so on.

Other effects were also taken into account, such as increased plant growth and decay, and the loss of soil moisture, but again these were

[14] Venus's atmosphere is 96% CO_2 and with a pressure of 90 times that of the Earth's. Under such conditions the two infrared bands of CO_2 experience pressure broadening, enabling tham to trap a wider range of radiant heat than on the Earth.

assumed only to make matters worse. All these factors were duly allowed for in the IPCC's model of the Earth's atmosphere, using a program referred to as the General Circulation Model (GCM). The output from this computer simulation was the foundation on which their predictions were based. The 1990 report also contained a rather misleading diagram that was to be reproduced many times. This showed that greenhouse gases produced from human activity contributed to the change in global warming in the proportions: CO_2 55%, CFCs 24%, methane 15%, and nitrous oxide 6%. It was not surprising that people began to read and hear that CO_2 was the single most important greenhouse gas.

Perhaps aware of what was to follow, the IPCC belatedly tried to maintain its credibility within the scientific community by stressing the uncertainty inherent in its predictions, but these reservations were not given quite the same publicity as their earlier warning of calamitous global warming. The IPCC acknowledges the difficulty in measuring global average temperature before the mid-19th century, but it felt able to say that since 1860 there has been an increase of 0.4 Celsius, mainly from 1910 to 1940. At the end of its first report in 1990 IPCC stressed the need for more funding so that researchers could have access to the fastest possible computers. This would enable them to take into account the links between the atmosphere and the oceans, and also to make more detailed regional predictions. The funding was forthcoming.

In 1992 the IPCC published a Supplement, the text of which was agreed at a meeting of Working Group I at Guangzhou, China, in January 1992, which was attended by 170 delegates from 47 countries. New modelling studies had not confirmed the wilder claims of the 1990 report. The IPCC now came to the conclusion that global warming due to CO_2 was not likely to be a runaway process after all, would not exceed 4.5 °C, and may even be as little as 1 °C. The IPCC also said that the increase in global warming in the past 100 years fell within the usual climate variability and could even be due to other causes. Their most telling statement was that 'the unequivocal detection of the enhanced greenhouse effect from observations is not likely for a decade or more.' The computer model was now able to take into account ocean–atmosphere effects but still could not cope with clouds, which of course have a profound effect.[15]

Moreover, a new uncertainty had crept into the computer modelling: sulfate dust. Specks of sulfate from industrial emissions of sulfur compounds have a cooling effect, especially in the industrialised Northern Hemisphere. The link between global temperature and the 11-year cycle of activity of the Sun was acknowledged as likely to play a part, but it was not possible to include this factor in the model. And the IPCC modellers could still not find where 7 billion tonnes of atmospheric CO_2 were disappearing to each year. They also calculated that most global warming is occurring at night time in the Northern Hemisphere—if it is occurring at all.

They also noted that the part of the greenhouse effect which could be attributed to CO_2 did not relate to the amounts of CO_2 in the atmosphere. Again to quote the 1992 IPCC report: 'the estimates of warming over the last 100 years due to increases in greenhouse gases made in the original report may be somewhat too rapid.' The report concludes that much more work needs to be done, especially in conjunction with social scientists, to develop new scenarios. What is needed, the report says, are better computer simulations, more accurate temperature measurements, more support for climate research around the world, and yet more meetings.

Clearly something had to be done, and alarmed by all this, governments around the world continue to pour funds into meteorological research. (The US Government alone will have spent $2 billion by the year 2000.) Global warming research groups sprang up with proposals on ways in which this bounteous funding could be employed. Most proposals were based on the use of larger computers, although some came up with ingenious but ingenuous ways of disposing of the CO_2 (see the Box on the next page).

The global warming facts, figures, fantasies and fictions are neatly summarised in *The Heated Debate* by Dr Robert Balling Jr, director of The Office of Climatology at Arizona State University, which was published in 1992. This readable and balanced account explains the background to the alarms of the 1980s. Balling pleads for a more

---

[15] We are all aware how important clouds are. The hottest days and the coldest nights are those without a cloud in the sky. Cloud cover moderates both extremes very effectively.

rigorous approach to the discussion of climatic research, and a more balanced approach to the forecasting of climate change. Both have been sadly lacking in recent years.

How to dispose of CO_2

Assuming that CO_2 emissions to the atmosphere should be reduced, then one way would be to burn more of the fuel which produces the least CO_2. Another way is to collect the CO_2 that is being generated and put it somewhere other than into the atmosphere.

Burning coal produces more CO_2 than burning oil, but burning natural gas produces least of all. A rough guide is that for every million calories of energy produced, we release about 400 grams of CO_2 if we use coal, 300 g if we use oil, and 200 g if we use natural gas. This solution to the problem would only last for a few years until the supplies of natural gas were exhausted, and this will happen anyway within 50 years, after which we will be forced to use more oil and coal. Oil will run out in about 125 years, although supplies of coal will last for over 1000 years.

Alternatively we can trap the CO_2 emitted by the major users of fossil fuels, such as power stations, and prevent its release to the atmosphere. One suggestion is that we should liquefy the gas and pump it into the deep oceans. The chemistry behind such a suggestion shows it is feasible. Collecting and shipping CO_2 as a liquid presents no technological problems, nor will pumping it to the ocean depths where the temperature is about 2 Celsius. At such a temperature a pressure of 37 atmosphere is needed to keep CO_2 a liquid, and this pressure is reached below 400 metres.

In theory, then, we can dispose of CO_2 into the ocean, and since there is little mixing between the depths and the surface layers, the CO_2 would be safely locked away. But just how much of the 22 billion tonnes of CO_2 which comes from fossil fuels each year we could reasonably dispose of in this way is another question—it would take 1000 pressurised supertankers each carrying 500 000 tonnes and making one trip a week to transport this amount. There is also a risk that a sudden upwelling of CO_2 from the depths might roll over the surface of the sea and suffocate all in its path, just as it did from Lake Nyos (see p 236).

Another more fanciful method of storing CO_2 has been proposed, and that is to store it on land as giant snowballs of solid CO_2 (dry ice). Walter Seifritz of the Institute of Energy Economics of Stuttgart came up with the amusing idea of snowballs the height of the Eiffel Tower and wrapped in glass wool to insulate them. A country like Germany would need to build two such snowballs a year, although to do so would consume half the energy produced in their power stations.

If disposal of CO_2 was deemed to be necessary then the most likely method of disposal would be to pump it back underground into exhausted gas and oil wells, and in the latter it might do a useful job by forcing more oil to the surface.

The good side of carbon dioxide

There are some ways in which carbon dioxide serves to improve the quality of our lives, and in some rather unexpected guises. Apart from its well-known ability to put the fizz into drinks, it will keep food fresher, it makes aromas smell more alluring, it can take a lot of the fat out of fried snacks, it can decaffeinate coffee, and it can be used as a gas for greenhouses. It can also be turned into other molecules such as solvents, plastics and painkillers.

Carbon dioxide appears in the top twenty industrial chemicals produced by the chemicals industry. The gas is collected as a by-product of ammonia manufacture and from the fermentation of alcohol. In the USA over a million tons of CO_2 are collected each year and used as a resource for the chemicals and other industries. In the UK it is 40 000 tons. As a starting material for other chemicals, the industrial use of CO_2 is somewhat limited. A little is reacted with phenol to make aspirin. Much more goes into manufacturing poly-carbonate plastics, and even more into the manufacture of **urea**. The CO_2 can be converted to urea by reaction with ammonia, and this white solid can be used directly as a fertiliser, as an animal food supplement, or reacted further and turned into plastics. Some CO_2 can be turned into the solvent propylene carbonate, which is used in the polymer industry.

CO_2 is now used to boost greenhouse crops, and this is done simply by pumping CO_2 into glasshouses. If you tried to grow plants under glass without any ventilation they would quickly stop growing because they would use up all the CO_2. With 1000 ppm of CO_2 in the air it is possible to increase crop yields by 40%. Plants grow bigger where there is more CO_2, and this is also true in the wild. Research by Professor James Teeri of the University of

Michigan has demonstrated that if the amount of CO_2 in the atmosphere were to double, then tree growth would go up by 20%.

CO_2 is increasingly used as a solvent with remarkable properties. Liquid CO_2 has the ability to dissolve a large range of materials, and especially organic compounds, such as fats and oils. Unlike most industrial gases, CO_2 is not transported in heavy cylinders, but either as a solid, referred to as 'dry ice', or as a liquid. Solid CO_2 was formerly used in long-distance transportation of food before refrigerated trucks were developed. It is still used to cool processed food, and a blast of CO_2 snow will ensure the rapid refrigeration of frozen meals. It is delivered to food processors as a liquid, in heavy-gauge steel tankers carrying up to twenty thousand litres at a time, and is converted to CO_2 snow on site. Tankers carry their cargo at pressures of about 70 atmospheres.

At atmospheric pressure CO_2 can never be a liquid, and if we cool it we find that at -78 °C it freezes into a solid. On warming it reverts straight back to a gas. This physical process is called sublimation. To get CO_2 to behave as a liquid we need to apply pressure. The higher the temperature the higher the pressure needed to keep it in the liquid state. It is possible to have liquid CO_2 at room temperature, and even up to 31 °C, when a pressure of 73 atmospheres is needed. This is called the critical temperature for CO_2, because above this temperature no pressure, however great, can keep CO_2 a liquid. Yet in industry and research, chemists work with CO_2 up to 100 °C and 300 atmospheres. Under these conditions CO_2 has both the properties of a liquid and of a gas, and is correctly referred to as a supercritical fluid.

Carbon dioxide is ideal for processing materials which eventually come into contact with humans, because it is odourless and tasteless, it does not leave any unwanted residues, and, unlike many solvents, it is not flammable. Industry finds it attractive because liquid CO_2 is cheap and chemically unreactive, it has a low viscosity, and it will dissolve a wide range of materials such as oils, fats, caffeine, perfume essences and other plant extracts. Extracting the essential oils of plants, herbs, flowers and spices using supercritical CO_2 produces much improved essences, which reflect the natural composition more closely than extracts using traditional solvents. Moreover, liquid CO_2 has no smell of its own, unlike other solvents which can leave residues.

Many of the fragrances discussed in Chapter 1 can now be extracted using liquid CO_2. Hops have been extracted in this way since 1977 for use in making beers.

Other oils that can be extracted with liquid CO_2 are the edible oils from fried foods. Surface oil on potato snacks can be washed off to produce the 'reduced fat' varieties, which have 25% fat as compared to untreated snacks with 40% fat. Liquid CO_2 is the solvent of choice for decaffeinating coffee. Earlier methods of making this used organo-chlorine solvents such as chloroform, but these left behind minute residues of solvent which still worried people. The best-selling dec-affeinated coffees, such as Café Hag in Europe, are made using liquid CO_2. The process involves extracting the caffeine from the beans with superheated water, and then extracting the caffeine from the water with supercritical CO_2. The result is to reduce the amount of caffeine in a cup of instant coffee from 60 mg to only 3 mg. Tea can also be decaffeinated, but this is a more complicated process since there are subtle tea flavours extracted along with the caffeine. However these can be recovered and put back.

The most recent suggestion for using liquid CO_2 is as a solvent for dyeing. The effluent from textile dyeing works is particularly difficult to clean up, and although the amount of chemical waste may be relatively small it certainly looks like a great deal. Some dyes are very soluble in liquid CO_2 and this turns out to be an ideal dyeing medium. There is no water involved and the results are often much brighter shades of colour. If we are prepared to invest in new tech-nology we may find that CO_2 is one of the chemicals that in the next century is actually *protecting* our environment.

One day we may use CO_2 to *make* other fuels such as methane, methanol, and even petrol. This can easily be done provided we have a supply of hydrogen gas to convert the CO_2 to methane (CH_4). We could get the hydrogen by the electrolysis of water from power generated by windmills, hydroelectricity, tidal barrages or nuclear power. We might even use sunlight to split water directly into oxy-gen and hydrogen, a process that chemists have shown is feasible but not very efficient. When sunlight shines on **titanium dioxide** parti-cles in water it will work this little miracle, provided the titanium dioxide has a small amount of platinum metal incorporated into its surface. Currently we do not use hydrogen to turn CO_2 into methane: in our topsy-turvy world we convert methane to CO_2 and hydrogen. The hydrogen is used as a chemical resource to make ammonia for fertilisers and plastics.

In our everyday lives the most pleasant way in which we come into contact with CO_2 is in colas, sparkling wines, mineral waters and fizzy fruit drinks. Traditionally wine-makers added more yeast, and got the sparkle by letting the CO_2 build up naturally in the bottle. Today it is added from industrial sources, but it is no different, although there are some who say that 'natural' CO_2 in expensive champagne or Perrier water produces finer bubbles, and that this makes for a better drink.

Another way in which CO_2 will improve drink is in the treatment of municipal water supplies. Water can be clarified by adding aluminium salts which precipitate and carry down impurities, so making it crystal clear. This use of aluminium is not wanted by some people, who consider this metal to be toxic and harmful to health. The alternative way of treating drinking water is to make it slightly acidic with CO_2 and then add iron salts. This use of CO_2 is likely to grow in the next century.

We have focused in this chapter on the CO_2 gas which comes from fuel and ends up in the atmosphere, but for each molecule of CO_2 that is formed, there is a molecule of oxygen consumed. Even though CO_2 may not be the global threat we have been led to believe, could it be that we are overlooking another threat? Are we using up the Earth's oxygen? The answer is yes. The burning of 7 billion tonnes of carbon from fossil fuels each year means we are consuming 18 billion tonnes of atmospheric oxygen.[16]

But there is no need to worry. The amount of oxygen we are losing each year is so small as to be almost unmeasurable, because it is trivial compared to the total amount of oxygen that surrounds the Earth: there are a thousand *trillion* tonnes circling the globe. It would take over 2000 years, even at the present rate of depletion, for the oxygen level in the atmosphere to fall a mere 1%, from 21% to 20%. Even if we were to consume all 16 trillion tonnes of accessible fossil fuel

---

[16] The 100 million tonnes of O_2 that industry extracts from the atmosphere, by condensing and by distilling liquefied air, is trivial in comparison. This goes to producing steel, making ethylene oxide (which is turned into antifreeze or polyester for bottles and fabrics) or used as oxygen gas itself, in medical care, or to purify sewage.

(mainly coal) in one glorious conflagration we would only use up 4% of the planet's oxygen, although this would put us dangerously close to the 17% minimum that is required to support human life.

Take care

TODAY the word 'chemical' is used by many people to mean pollutant, unnatural, or dangerous. It once simply referred to something that was made and used by chemists. The odium which now surrounds the word 'chemical' is a symptom of a science in decline. Why has this happened to chemistry? When I was young, chemistry needed no one to defend it—its achievements spoke for themselves. In the 1960s, much of what was reported about chemicals in the media was positive, and this encouraged me to believe that my choice of subject to study at university had been the right one. Then in the 1970s and 1980s what I read became increasingly negative, until the very word 'chemical' became synonymous with danger, and was only used in that context. Like many of my fellow scientists I objected to this, especially so when what was being said about 'chemicals' was in fact wrong.

Chemists are puzzled by the way some people have a strong aversion to *all* things 'chemical', and there is even a word, chemiphobia, to describe this state of mind. Chemiphobia stems not from a fear of industrial accidents, but from the dangers which people see lurking in everyday materials. Nevertheless, our public image was not helped by a sequence of terrible chemical disasters in the 1970s and 1980s, such as the mass poisoning of the people of Bhopal, India, in 1984, or the killing of all the fish in the river Rhine in 1986. Tighter safety precautions followed, but the damage was done. The chemicals industry was to pay for appearing to be so uncaring; and because it was such a profitable industry there were those who extorted a heavy price.

In this respect it seems rather strange that other industries which inflict such damage on the public have suffered much less, and we rarely see protestors attacking the work places of car makers, tobacco companies and arms manufacturers. Their products kill and maim far more often than the chemicals industry. I find it hard to understand. Thousands of my fellow chemists work away in research laboratories, and with the best of motives, to make better healing drugs, stronger plastics, safer solvents, better cleaners, and tastier germ-free food. It is thanks to them that those same protestors will live richer and longer lives, and even enjoy a pain-free old age. But no matter how hard chemists toil, their public image seems fated not to improve.

It is not the chemicals industry as such that has led to chemiphobia, but the many alarms about 'chemicals' in everyday products which we are told could possibly be damaging our health or the environment. Additives, pollution, residues and side-effects are the issues that have led some people to question the use of 'chemicals' and to advocate 'natural' alternatives. It is just these areas that I have sought to tackle head on in *The Consumer's Good Chemical Guide*, and I hope that you will now see artificial sweeteners, dioxins, carbon dioxide, PVC and painkillers in a slightly less worrying light.

I have also looked at other areas where a little chemical understanding can make the issue seem less frightening than some would claim, and in this context I have dealt with aspects of our diet such as sugar, cholesterol, saturated fats and nitrates. And just to show that chemists are involved in areas that are not generally regarded as their province I have written about their role in making better perfumes and safeguarding our favourite relaxant, alcohol.

Chemists are saddened by the fact that our language has been hijacked and turned to our disadvantage—the words 'chemical' and 'drug' are the examples they cite most often. For many people 'drug' has become a four-letter word meaning a drug of abuse or addiction, as in the cruelly dismissive phrase 'drug addict'. Yet the word also means just the opposite, a drug that heals—and this was how it was once used. Most people do not confuse these different meanings, even though they do overlap, as we saw in Chapter 5.

The misuse of the word 'chemicals' is more worrying. We live in a world in which the ultimate cause of things can generally be traced to a chemical process, whether these be the shape of a snowflake, the smell of freshly ground coffee, the colour of blood or the warming of the Earth. The same is true of the human body, and when things go wrong with that we may look around for a chemical cause. If the threat of heart disease or cancer, the two grim reapers that eventually mow most of us down, can be linked to 'chemicals' we become alarmed. Yet chemicals are rarely the cause of disease, and the major factors remain the ones that have always been important: a hereditary tendency to heart disease or cancer; poor diet and lifestyle leading to overweight; lack of exercise; and smoking. Those who fear these illnesses and feel at risk are perhaps particularly open to the suggestion that the fault lies not within themselves or what they do, but in some hidden threat.

Because we find it hard to come to terms with death, we are particularly sensitive to things which threaten to bring it about prematurely, and yet we are illogical in our fears. It is no good my telling you that you are a million times more likely to die in a car accident than from drinking artificially sweetened drinks. You will find people taking action to remove the latter threat from their lives, while making no effort at all to change the way they drive.

Perhaps it is natural to look for an external threat, because we find it difficult to accept that internal ones are beyond our control. Thus we may be predisposed to believe someone who tells us that it is sugar or fats that cause heart disease, that PVC causes cancer, that dioxins lead to deformed babies, and that nitrates can kill. The language of those who make these charges is carefully framed, since the charges are not proven. They use carefully crafted phrases like: 'evidence suggests that', or 'environmentalists believe that', or 'medical data point to'. They speak the language of doubt while appearing to be sure. They use the words of uncertainty but with conviction, and say ' it is believed that ...', or 'indications are that ...' They talk of 'facts that are connected...' or 'factors which appear to be linked ...' Phrases like these should put us on our guard, because they warn us that there is no *scientific* proof for what is being suggested, and we would do well to question those who are making the claims.

If you ask a group of people about their health and behaviour you can do lots of statistical analysis on such a survey, and maybe even discover some relationships that at first sight are not obvious. This analysis may reveal that people in certain areas are more likely to suffer stomach cancer, and you may then look for things in their food or water supply to explain this. Given the recent dramatic advances in chemical analysis, you might discover there are things in our environment or diet of which we were previously unaware, and the temptation will be to fix on these 'new' threats for an explanation. Indeed, our current awareness of environmental issues rests on the achievements of chemical analysts, who can now detect and measure some substances that are present in only *trillionths* of a gram. These advances in chemical analysis rarely hit the headlines; but what is discovered often makes the news.

It is easy for someone to accuse a particular chemical of causing illness, but much more difficult for the chemist to prove that it is innocent. And even when this is done it will rarely be given the same coverage in the media as the original charge. This book has been an attempt to try and level the playing field a little. I have taken a different view of chemicals, not from the top down, but from the bottom up. I did not start with alarming statistics about heart disease or cancer, and possible links to unsuspected chemical causes; instead I tried to explain the chemistry and then see if this is consistent with the accusations being made against the chemical. This way we discovered that some of the most famous scares of the past few years had little substance, and that some of the 'chemicals' that people worry about pose no danger to their well-being at all. There are still many threats to our health, our sanity, our society, our world, and our children's future, but these threats are not to be found in 'chemicals'. In the next volume of *The Consumer's Good Chemical Guide* I hope to deal with other chemicals that have given rise to alarm, such as phosphates, MSG, preservatives, chlorine, nicotine, methanol and the HFCs, and some that have led to unrealistic hopes of improved health, such as vitamin C and calcium. I will also look at topics that reveal a more positive side to 'chemicals', such as colour.

Whenever a scare story starts, you trust that the right questions will be asked of those who are banging the warning drum. Up to now this questioning rarely seems to have happened, but then the alarmists are rarely cross-examined by a science-trained interviewer. If it is a health scare, as indeed most of them are, then they should be asked the following questions: 'Have you compared those affected with a control group?' 'Has the harm that you say is being done been confirmed by doctors?' 'Have you repeated your study and got the same results each time?' The answers to these questions should be our first line of defence if we want to be sure of the truth. Unfortunately, they rarely seem to be asked.

If you want to put the risks associated with 'chemicals' and 'drugs' into their proper context, I would advise you to read one of the books on risks listed in the Bibliography, such as *Living with Risk* by Michael Henderson. These will explain the risks we run in all walks of life. *Living with Risk*, for example, discusses the main causes of death, and the factors which have the most influence on our health. Read it and you will discover how tiny is the threat from chemicals.

Finally, a word to the members of my own profession, the chemists. We are now resigned to seeing the things we produce misrepresented as dangerous, even though we have tested them and proved that they are safe. Even we tend to forget that chemicals have saved more lives than surgeons, and created more wealth than bankers. Of course we have made mistakes, and this book has reminded you of some of them, but for every mistake there have been many more successes. The greatest danger for the future is that young people are turning away from chemistry. We should do all we can actively to promote our science. The time has come for us to come out of the bunker and fight back. We know that remarkable things can be achieved by chemistry, but only if we capture the best young brains of each generation and direct them to a career in research in the laboratory, or to manufacturing new materials and drugs in the chemicals and pharmaceuticals industries. Once upon a time this happened, and looking around us we can see how the achievements of that earlier generation of chemists changed the world. For the sake of the world in the next century, we must take care it happens again.

Chemical Data

The Consumer's Good Chemical Guide is all about chemicals, but not about their chemistry as such. If you are seeking that kind of information then here are some of the bare essentials. Compounds are listed in alphabetial order according to the common name used in the main body of the text. The formula is given next, followed by the other information; the molecular structure appears in the margin. In many instances I have given alternative names for the various chemicals mentioned, including the chemical name. You will need this if you want to look for further information by consulting the general chemistry reference works listed at the end of this Appendix. To find a particular chemical, ignore any single letter such as l-, d- or *N*- at the start of a word, and any numbers and Greek symbols. Acronyms such as PVC are in alphabetical order.

Where a zig-zag line is used this represents a chain of CH_2 groups, e.g. stearic acid on page 319.

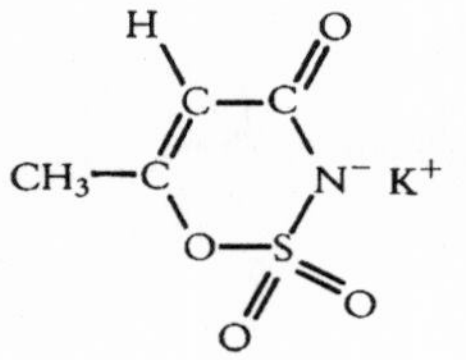

acesulfam: $C_4H_4KNO_4S$, is the intensely sweet-tasting potassium salt of 6-methyl-1,2,3-oxathiazin-4-one 2,2-dioxide. It is also called acesulfam-K when used as an artificial sweetener.

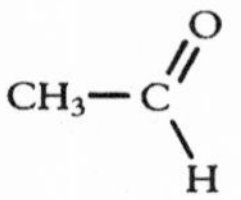

acetaldehyde: C_2H_4O, more correctly called ethanal. It is a colourless liquid with a pungent, fruity odour which boils at 2 °C. Normally it is available as a solution in water, and used as such. Acetaldehyde used to be produced on a large scale industrially since it is the forerunner of many other chemicals such as acetic acid and acetic anhydride, but these are now made by alternative methods.

acetanilide: C_8H_9NO, can be obtained as white, shiny, crystals which melt at 114 °C. It is used as a stabiliser for hydrogen peroxide and cellulose coatings. It is also used in the manufacture of pharmaceuticals, dyestuffs and antiseptics. (See **paracetamol**.)

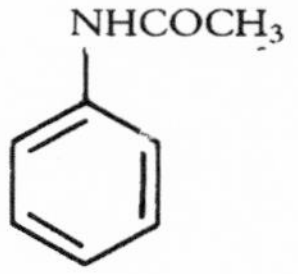

acetate: the salt component of **acetic acid**.

acetic acid: $C_2H_4O_2$, more correctly called ethanoic acid. It is more popularly known in the form of vinegar which is a dilute solution of the acid in water. Pure acetic acid is a colourless, sharp-smelling liquid which boils at 118 °C and is produced industrially. The acetic acid in vinegar is manufactured by the action of bacteria on alcohol and the name vinegar comes from the Frnech *vin egar* meaning sour wine.

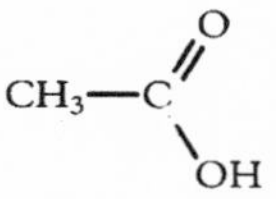

acetone: C_3H_6O, also known as dimethyl ketone, but is more correctly called propanone. Acetone is a volatile liquid which boils at 56 °C and is used widely as a solvent, both industrially and also domestically for such things as nail varnish remover.

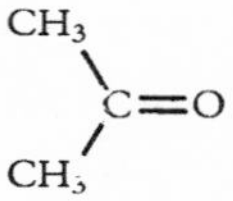

acetylsalicyclic acid: See **aspirin**

N-acetylcysteine: $C_5H_9NO_3S$, the N-acetyl derivative of the naturally occurring amino acid, cysteine. It is used in medicine and biochemical research.

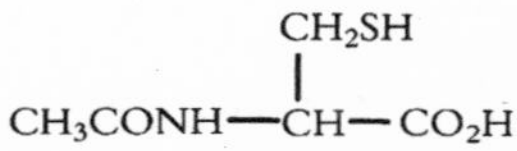

acids: see **carboxylic acids**

acrolein: C_3H_4O more correctly called 2-propenal. It is a colourless or yellow liquid which boils at 53 °C and has a disagreeable choking odour. It is produced when organic materials and plastics catch alight, and is a major cause of death of those suffocated in house fires. Acrolein is produced as an industrial chemical for the manufacture of plastics, resins, pharmaceuticals and herbicides.

acrylic acid: $C_3H_4O_2$, also more correctly known as propenoic acid. This is a colourless liquid which boils at 141 °C and has an acrid odour. It polymerises readily, and so does its esters which are used in surface coatings, emulsion paints and to treat paper and leather.

$$CH_2{=}CH{-}CO_2H$$

acrylonitrile: C_3H_3N, also known as vinyl cyanide or more correctly, propenenitrile. It is produced industrially on a large scale by the reaction of propylene, $CH_2{=}CH{-}CH_3$, ammonia and air at 400 °C in the presence of a molybdenum-based catalyst. Most acrylonitrile is polymerised to make fibres and plastics.

$$CH_2{=}CH{-}CN$$

alcohol: the general name for a class of chemical compounds which have a hydroxy group (OH) attached to a carbon atom. The simplest alcohol is methanol, CH_3OH, then ethanol, propanol, butanol etc. (See individual entries.)

aldehyde: the general name for the group CHO, and for the chemical compounds in which this group occurs. Alcohols are oxidised to aldehydes, which in turn are oxidised to acids, thus:
$$RCH_2OH \rightarrow RCHO \rightarrow RCO_2H$$
(R = an organic group)

alginates: any of several salts of alginic acid which is a polysaccharide consisting of long chains of joined units of mannuronic acid and guluronic acid. Alginates occur in all brown algae as a component of their cell walls, and

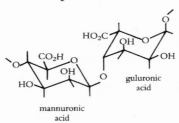

mannuronic
acid

guluronic
acid

the major source is the giant seaweed *Macrocystis pyrifera*. Alginates make excellent thickening agents and are used in many foods such as ice cream, salad dressing, and milk chocolate. They are also used to stabilise fresh fruit juice and the foam on beer.

alkane: general formula C_nH_{2n+2}, the name given to a series of hydrocarbons starting with the simplest, methane, CH_4, which has only one carbon atom, then going up through ethane, C_2H_6, propane, C_3H_8 and butane, C_4H_{10}, after which they are named after the corresponding Greek numbers: pentane, C_5H_{12} hexane, C_6H_{14}, heptane, C_7H_{16}, and so on. Some alkanes are given as separate entries in this Appendix. The early members of this series are gases, then come volatile liquids whose boiling points increase with increasing molecular weights. Eventually when there are many carbons we have thick oils, greases and finally waxes. In everyday life we encounter these hydrocarbons not as pure compounds but as mixtures which are called gasoline/petrol, diesel, kerosene/paraffin, white spirit and liquid paraffin. The old name for the alkanes was the paraffins.

CH_3—CH_3
ethane

CH_3—CH_2—CH_3
propane

etc.

allyl alcohol: C_3H_6O, more correctly known as 2-propen-1-ol, and is a colourless liquid with a pungent smell and boiling point of 97 °C. It is produced industrially on a large scale as the starting material for resins and plasticisers, and as an intermediate in the production of other chemicals.

H_2C=CH—CH_2OH

ambergris: $C_{16}H_{29}O$, the main aroma constituent of the balls of ambergris produced in the intestines of sperm whales.

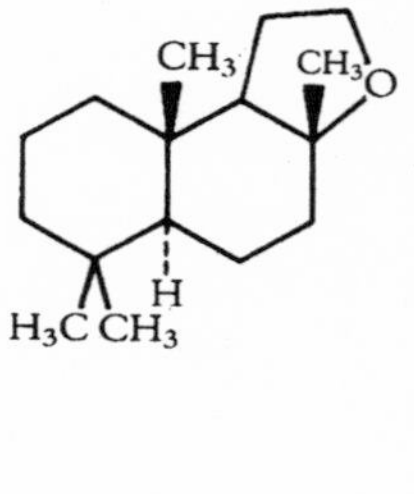

amino acids: a collection of chemical compounds with the common grouping shown where R stands for a variety of other atoms and groups. The simplest amino acid is glycine

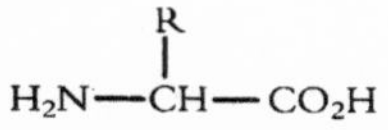

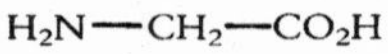

$$H_2N—\underset{|}{\overset{R}{C}}H—CO_2H$$

$$H_2N—CH_2—CO_2H$$

glycine

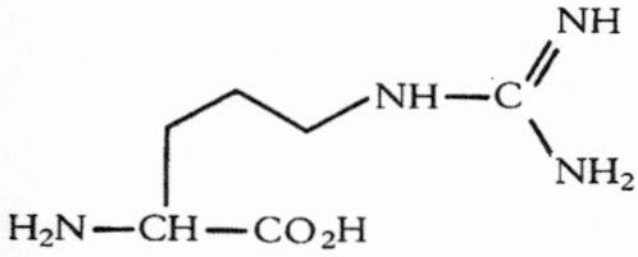

$$H_2N—CH—CO_2H$$

arginine

(coded gly) in which R is hydrogen. There are over 20 amino acids found in living things, some of which are essential to humans, such as **arginine** (arg) which has $R = NH_2C(NH)NH(CH_2)_3$. The other essential amino acids are histidine (his), iso-leucine (ileu), leucine (leu), lysine (lys), methionine (met), phenylalanine (phe), threo-nine (thr), tryptophan (trp) and valine (val). Amino acids join together by reacting the amino group (NH_2) on one with the acid group (CO_2H) on another to form a peptide bond ($-CHR-CO-NH-$etc). Long chains are known as peptides, and very long chains as polypeptides. Proteins are polypeptides. Glutamic acid (glu), better known as its sodium salt, mono-sodium glutamate (MSG), is also an amino acid in which R is $CH_2CH_2CO_2H$.

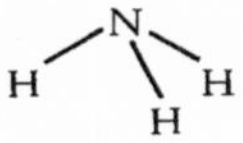

ammonia: NH_3, a colourless gas with a strong odour. It condenses to a liquid at $-33\ °C$ which has the unique ability to dissolve metals like sodium without reacting with them. Ammonia is manufactured on a large scale industrially and used to make fertilisers, plastics and nylon. Ammonia gas dissolves readily in water to give a solution popularly known just as 'ammonia'.

ammonium nitrate: NH_4NO_3, a salt of nitric acid which is formed by reacting the acid with ammonia. It is a colourless, crystal-line material which decomposes on heating to form nitrous oxide gas, N_2O, and water. Solid ammonium nitrate tends to cake into lumps and is treated with anti-caking powders to keep it free flowing. It is produced on a large scale worldwide, principally as a fertiliser, although some is used as an explosive when mixed with other materials.

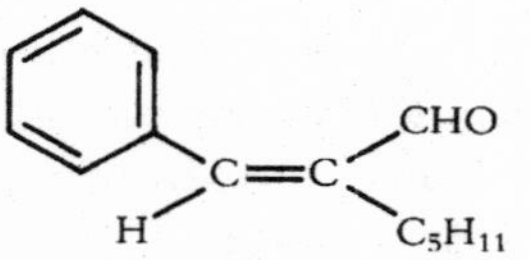

α-amyl-cinnamaldehyde

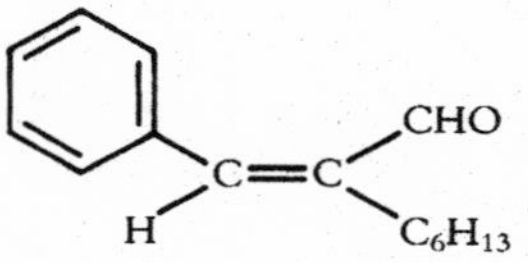

α-hexyl-cinnamaldehyde

α-amyl-cinnamaldehyde, $C_{14}H_{18}O$, and **α-hexyl-cinnamaldehyde, $C_{15}H_{20}O$,** are both yellow liquids which can be synthesised from benzaldehyde. They have soft, floral odours with distinct jasmin notes.

androstenone: $C_{19}H_{28}O$, also known as 5α-androstenone, and more informatively as 16-androsten-3-one. It is a molecule consisting of four joined rings in an arrangement which is characteristic of a steroid. Its melting point is 103 °C. Androstenone is responsible for the smell referred to as 'boar taint' in pig meat.

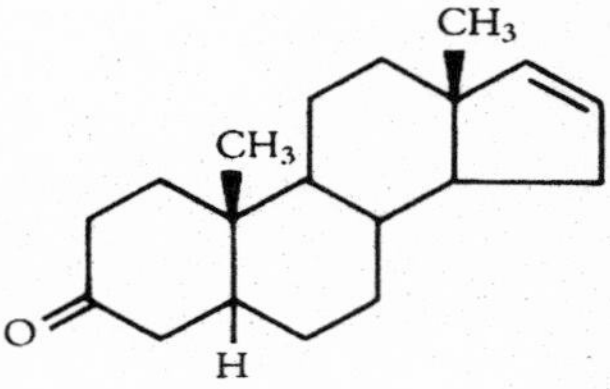

Antabuse: the trade name for disulfiram, $C_{10}H_{20}N_2S_4$, more correctly called bis(diethylthiocarbamyl) disulfide. It can be given as part of the treatment of alcohol addiction, and it inhibits the enzyme aldehyde dehydrogenase. If alcohol is taken during treatment with the drug, which is prescribed as 200 mg tablets, then the patient will suffer unpleasant effects such as nausea and vomiting. One of its side effects is to produce bad breath, because of its high sulfur content.

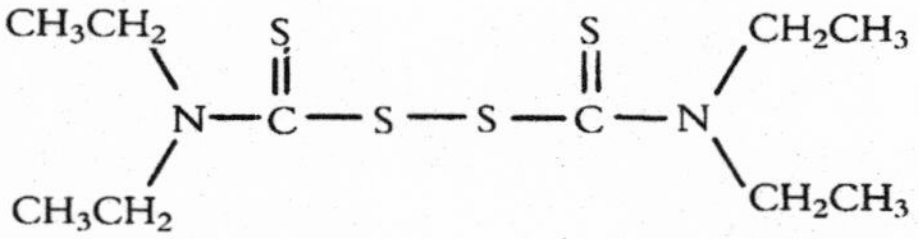

antimony oxide: Sb_2O_3, also more correctly known as antimony trioxide. It is found as the ore valentinite, or it can be made by burning antimony in air. It is used to flameproof textiles, paper and plastics, especially **PVC**. It is also used to decolourise glass.

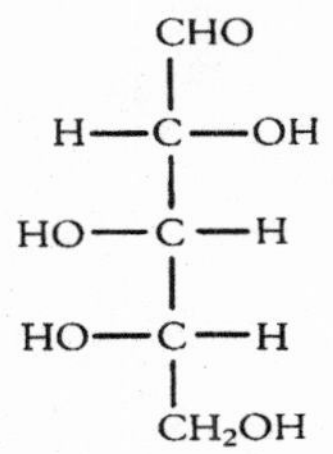

```
        CHO
         |
  H —— C —— OH
         |
  HO —— C —— H
         |
  HO —— C —— H
         |
        CH₂OH
```

arabinose: $C_5H_{10}O_5$, also known as pectin sugar and gum sugar. Both the left and right hand forms of this sugar occur naturally and L-arabinose is common in vegetable gums, of which gum arabic is the richest source. Arabinose is a white crystalline material, which is soluble in water, and melts at 159 °C. It is used as a culture medium.

L-arginine: see amino acids.

argon: element, atomic number 18, atomic mass 40, symbol Ar. This element is completely unreactive and as such exists in the air only as a gas composed of single atoms. It condenses to a liquid at −186 °C. Argon is 1% of the atmosphere, totalling 66 trillion tonnes (66×10^{12}). It is so abundant because it is being continually formed by the radioactive decay of one of the isotopes of potassium (^{40}K). About 700 000 tonnes are extracted each year from the air for lamps and high temperature metallurgy, where it is used to provide an inert atmosphere.

arsenic: element, atomic number 33, atomic mass 75, symbol As. Arsenic is used in alloys, semiconductors, glass and pesticides. The term 'arsenic' in literature refers to the oxide, As_2O_3, a white powder which is slightly soluble in water and acts as a deadly poison. The element occurs widely and is essential for some species including humans. Our daily diet provides between 5 and 50 mg of arsenic, and it is high in seafoods like plaice and prawns. We excrete arsenic rapidly. A small amount of arsenic can act as a tonic by speeding up our metabolic rate, and it has been used to rear livestock more quickly. However, too much arsenic over a long period can cause cancer.

asbestos: the name given to a family of fibrous minerals that are impure magnesium silicate. Their colours vary and some have individual names such as chrysotile, whose fibres are strong enough to be woven, and amphibole, which is very resistant to heat and chemical attack. Asbestos was once used for fireproof fabrics, brake linings, building panels, roofing components, rubber reinforcement, and chemical filters. It is still used in the diaphragm cells for chlorine manufacture. Exposure to asbestos dust in industry caused workers to be affected with the lung disease asbestosis and cancer. Its use is now strictly controlled.

aspartame: $C_{14}H_{18}N_2O_5$, (brand name NutraSweet) a white crystalline powder slightly soluble in water and alcohol.

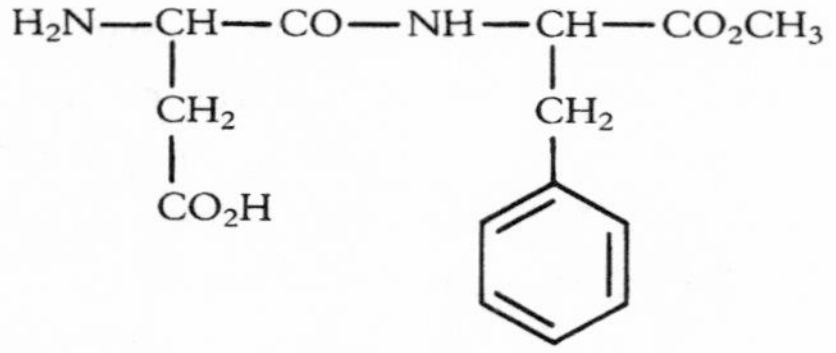

$$H_2N-CH-CO-NH-CH-CO_2CH_3$$

aspartic acid: $C_4H_7NO_4$, a naturally occurring but non-essential amino acid. The common form is L-aspartic acid which can be obtained as colourless crystals from young sugarcane or sugar beet molasses.

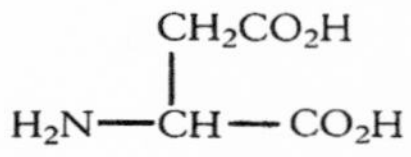

$$CH_2CO_2H$$
$$H_2N-CH-CO_2H$$

aspirin: also known chemically as acetylsalicylic acid. It is a white crystalline material with a slightly bitter taste, and it melts at $132\,°C$. It is stable in dry air but in moist air it slowly breaks up into salicylic acid and acetic acid.

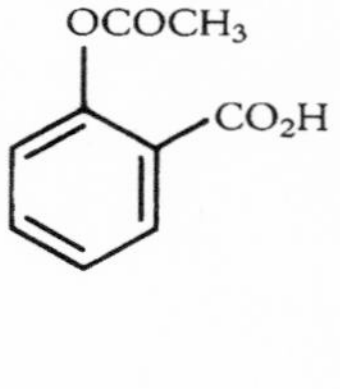

auranofin: the correct chemical name for this anti-arthritic drug is 2,3,4,6-tetra-*O*-acetyl-1-thio-D-glucopyranosato-*S*-(triethyl-phosphine)gold.

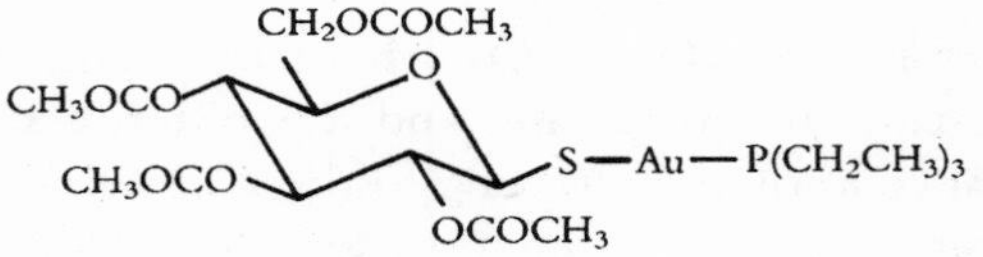

$$CH_2OCOCH_3$$
$$CH_3OCO$$
$$CH_3OCO$$
$$S-Au-P(CH_2CH_3)_3$$
$$OCOCH_3$$

Bakelite: a trade name for a whole range of cross-linked co-polymers, made from phenol and formaldehyde. Leo Baekeland (1863–1944) discovered that when these chemicals reacted, and the product was heated under pressure, it turned into a useful resin which he called Bakelite. In 1910 he established a company to make products of the new plastic which became very popular, and he became very wealthy.

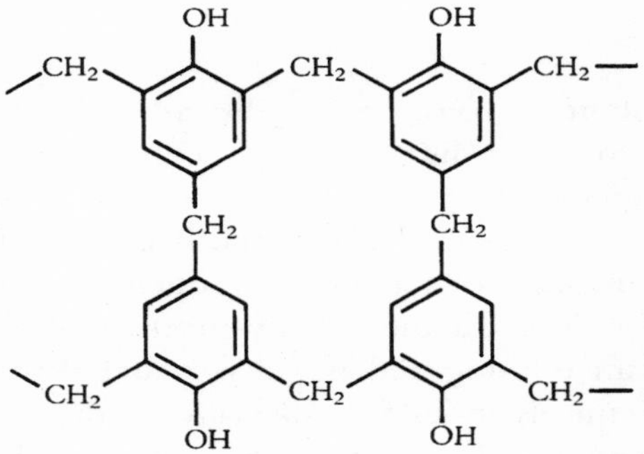

OH

CH₂ — CH₂ — CH₂ —

CH₂ — CH₂

CH₂ — CH₂ — CH₂ —

OH   OH

benzene: C_6H_6, the simplest of the so-called aromatic hydrocarbons. It is produced industrially on a large scale because it is the feedstock for many processes and is used in the manufacture of a large number of chemicals that end up as polymers, plastics, detergents, drugs and dyes. Benzene itself is a colourless, volatile liquid which boils at 80 °C. It is easily ignited and burns with a smoky flame. It was commonly used as a solvent in industry and dry-cleaning, but is now strictly controlled because exposure to its vapour over a long period can cause certain types of cancer.

Bitrex: $C_{28}H_{34}N_2O_3$, the trade name for denatonium benzoate, and the bitterest substance known, according to the *Guinness Book of Records*. It is a white powder that is added to household and garden chemicals to deter young children from drinking them. It is also used to denature industrial alcohol, which requires only 10 mg per litre (only 10 parts per million).

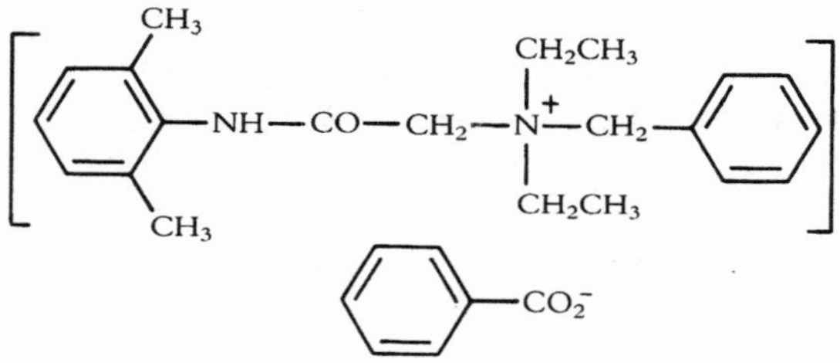

butane: C_4H_{10} is a hydrocarbon gas under normal conditions, but it condenses easily under pressure to a liquid which is sold as camping gas or lighter fuel. It is increasingly being used as an aerosol propellant in place of CFCs. Butane is a by-product of petroleum refining and is a raw material for the chemicals industry, especially in the manufacture of butadiene which is turned into synthetic rubber. There are two kind of butane molecule: n-butane (the n stands for normal) with its carbon atoms in a row, which condenses to a liquid that boils at $-0.5\,°C$; and *iso*-butane, which has a branched chain, and boils at $-12\,°C$.

$$CH_3-CH_2-CH_2-CH_3$$
n-butane

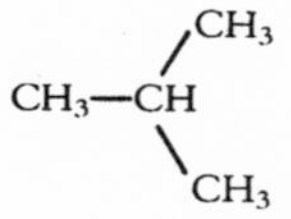

iso-butane

2,3-butanediol: $C_4H_{10}O_2$, also known by its older name of 2,3-butylene glycol. It is derived from starch and sugar beet molasses.

CH₃
|
CHOH
|
CHOH
|
CH₃

CH₃—CH₂—CH₂—CH₂OH

butan-1-ol

CH₃—CH₂—CH—CH₃
|
OH

butan-2-ol

CH₃
|
CH₃—C—CH₃
|
OH

tertiary butanol

2,3-Butanediol is a colourless crystalline material, but because it melts at 23 °C it is often found as a liquid. It boils at 179 °C and is soluble in water and alcohol. 2,3-Butanediol is used in resins and as a solvent for certain dyes.

butanol: $C_4H_{10}O$. There are three forms of butanol: butan-1-ol, which boils at 118 °C; butan-2-ol, which boils at 100 °C; and tertiary butanol (more correctly called 2-methylpropan-2-ol), which boils at 83 °C. The liquids are similar in physical properties, but in their chemical reactions they lead to very different products. They are used as intermediates in synthesising other chemicals.

caffeine: $C_8H_{10}N_4O_2$, also known as methyltheobromine and 1,3,7-trimethylxanthine. When pure it consists of white, long, flexible crystals which melt at 237 °C. It is soluble in chloroform, and is extracted from coffee beans or kola nuts. Caffeine is the active ingredient in coffee, tea and colas, and is used medicinally in the treatment of asthma. Its effect on the body is to stimulate metabolism. Despite claims to the contrary research has shown that it is not addictive.

calcium: element, atomic number 20, atomic mass 40, symbol Ca. It is a soft silvery metal that reacts slowly with air and water. It does not exist as such in Nature, where the only stable form is the calcium ion, Ca^{2+}, in which the calcium atom has lost two of its negatively charged electrons. This is also the form in which calcium occurs in food and in the body, and as such it is essential to all living things. The average person contains about a kilogram of calcium, most being in the skeleton as calcium phosphate. Rich sources of dietary calcium are sardines, dairy products such as milk and cheese, and white bread. Calcium is the fifth most abundant element in the Earth's crust, and there are vast deposits of **calcium carbonate** (limestone), **calcium phosphate** (apatite) and **calcium sulfate** (gypsum).

calcium carbonate: $CaCO_3$, commonly referred to as limestone. This sedimentary rock takes many forms, including marble. Calcium carbonate as a pure chemical is a colourless solid. It evolves carbon dioxide when heated strongly to give lime (CaO):

$$CaCO_3 \rightarrow CaO + CO_2$$

Calcium carbonate is used in smelting iron, and for manufacturing glass and cement. It is also used to absorb acidic gases, and to neutralise acid solutions (for example in indigestion tablets.)

calcium oxalate: CaC_2O_4, the calcium salt of oxalic acid. It is a white crystalline powder that finds some use in making glazes and in chemical research.

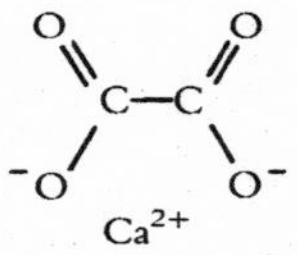

calcium phosphate: $Ca_3(PO_4)_2$. There are several types of calcium phosphate, but all consist of a lattice of positive calcium ions (Ca^{2+}) and negative phosphate ions (PO_4^{3-}). One form of calcium phosphate, called apatite, has the formula $Ca_5(PO_4)_3(OH)$ and is the main component of tooth enamel. Another, called fluoroapatite, has the formula $Ca_5(PO_4)_3F$ and occurs as vast mineral deposits.

camphor: $C_{10}H_{16}O$, a white solid with a characteristic medicinal smell. It occurs naturally in the roots and branches of the camphor tree, *Cinnamomum camphora*, which grows in China and Japan. It melts at 179 °C. It was previously much used in medicine, and as a plasticiser for cellulose nitrate, and in making lacquers.

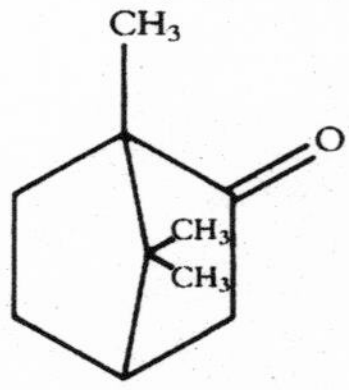

caramel: this is a food colourant of indeterminate chemical composition, made by melting and heating sugar. It can also be made from liquid corn syrup by heating it in the presence of catalysts at 120 °C for several hours. Caramels can be designed for use as either flavours or colourants. In the latter case the reaction is done in the presence of ammonia.

carbon dioxide: CO_2, a colourless, odourless, gas which is 1.5 times as dense as air. It is produced in large quantities as a by-product of fermentation or fertiliser production. When CO_2 is cooled it does not condense to a liquid but to a solid, popularly known as 'dry ice'. When a piece of this is warmed it sublimes at −78 °C, which means that it evaporates from the solid to the gas without going through a liquid form. CO_2 is soluble in water to the extent of 760 ml per litre, and this can be increased by raising the pressure when the product is called soda water or carbonated water. This is the basis of 'sparkling' mineral waters, colas, and other soft drinks.

$$O=C=O$$

carbon monoxide: CO, a colourless, odourless and deadly gas, discovered by Joseph Priestley in the USA in 1799. It is highly toxic because it binds to the haemoglobin in the blood 200 times more strongly than oxygen and thereby prevents red blood cells from carrying out their vital function. CO will condense to a liquid at −190 °C. It is produced on a large scale in industry by the action of steam on hot coke or natural gas, and is turned into methanol. CO is flammable, and can be used as a fuel—it burns with a pale violet flame.

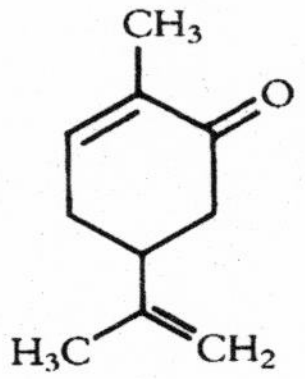

carvone: $C_{10}H_{14}O$, a pale yellow liquid which boils at 230 °C. Its correct chemical name is 2-methyl-5-(1-methylethenyl)-2-cyclohexene-1-one. The molecule is found in its mirror image forms, with D-carvone being present in caraway seeds and dill, while L-carvone is the flavour component of spearmint.

caustic soda: **NaOH**, often referred to simply as 'caustic' or by the old name of 'lye', but more correctly called sodium hydroxide. It is a white compound that becomes moist when exposed to the air. It is sold as a household chemical for unblocking drains. Caustic soda is produced on a large scale along with chlorine by the electrolysis of brine, and it is used industrially in the manufacture of rayon, pulp and paper, aluminium, soaps and detergents, textiles and vegetable oils. It is highly corrosive.

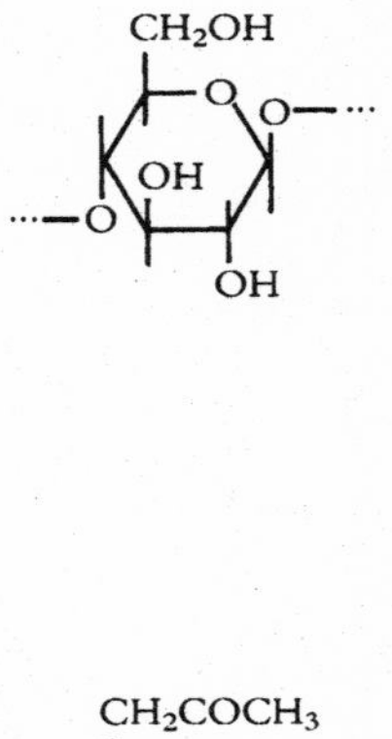

cellulose: $(C_6H_{10}O_5)_n$, a long chain carbohydrate polymer more correctly known as a polysaccharide. It consists of glucose units interconnected through oxygens. The cellulose from wood has about 1000 units while that from cotton has about 3500. Cellulose is the most abundant organic material in the world. It is a colourless solid which is insoluble in both water and organic solvents. As the raw material, its most important uses are in paper and clothing, but it is also chemically modified and turned into **cellulose nitrate** and **cellulose acetate**.

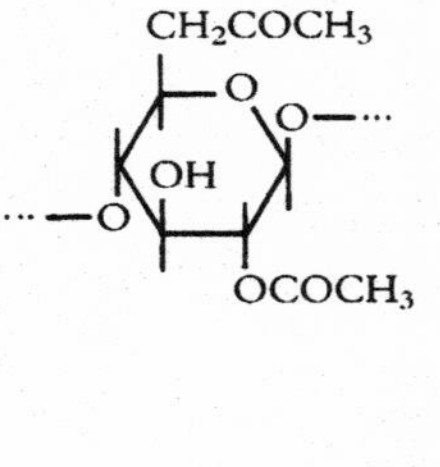

cellulose acetate: when **cellulose** from wood pulp or cotton is treated with acetic anhydride and acetic acid, it forms the acetate. Not all of the three replaceable hydrogens of a glucose unit of cellulose are replaced by acetate, only two of them on average. The product is a polymer that has a nice feel against the skin, and it has many outlets: fibre for clothing and furnishings; lacquer;

transparent sheeting (Cellophane); cigarette filters; magnetic tape; spectacle frames and screwdriver handles.

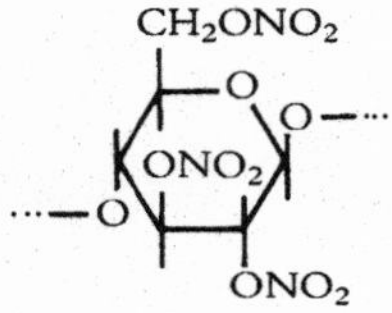

cellulose nitrate: also known as nitrocellulose or gun cotton. Replacing some or all of the reactive OH groups of the glucose rings of cellulose with nitrate groups, NO_3, gives cellulose nitrate, and its formula is approximately $C_6H_7N_3O_{11}$. It is made from cellulose using a mixture of nitric and sulfuric acids, and the product can vary from a pulpy solid to a liquid, depending upon the conditions of the reaction. Cellulose nitrate is highly flammable, but was widely used for fast-drying paints for car bodies, in explosives, and in Celluloid. Collodion was a solution of cellulose nitrate in ether and ethanol, and this was used in model making, for removing corns, and in engraving and photography. Because it is dangerously flammable it is no longer available.

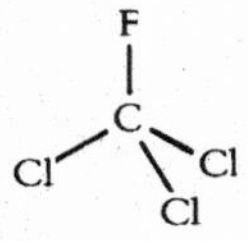

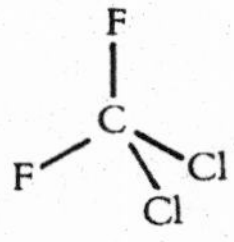

CFCs: this is the acronym for chlorofluorocarbons, which are compounds with one or two carbon atoms to which are bonded only chlorine or fluorine atoms. They were once used in aerosols and are still used in refrigerators, but they are now being phased out around the world because of the damage they can inflict on the ozone layer. The most important ones, and the ones still present in the atmosphere, are CFC-11 (which has the chemical formula $CFCl_3$), CFC-12 (CF_2Cl_2) and CFC-113 ($C_3F_3Cl_3$).

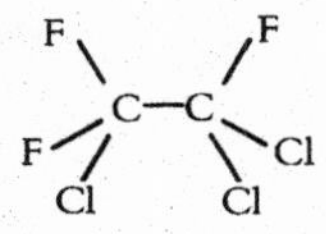

chlorine: element, atomic number 17, atomic mass 35.5, symbol Cl. The element occurs as a pale yellow-green gas of formula Cl_2. It is highly toxic. Chlorine gas can be condensed to a liquid either by applying pressure (8 atmospheres), or cooling to $-39\,°C$. Chlorine is produced on a large scale industrially and used for making plastics, especially PVC, and in water purification.

Many of its uses have been curtailed, such as the manufacture of **DDT**, **CFCs**, and solvents such as carbon tetrachloride. The manufacture of chlorine is carried out in a so-called chlorine cell. In this an electric current is passed through concentrated brine (sodium chloride solution) which releases chlorine at the positive electrode and sodium at the negative electrode. The chlorine that comes off has to be kept isolated from the sodium, and the older method of doing this was called the amalgam process which used liquid mercury as the negative electrode. This dissolves the sodium as it is formed to give a sodium–mercury amalgam, and this is run out of the cell to a separate compartment. There it comes in contact with water and immediately reacts to form **caustic soda**. Having given up its sodium the mercury is then pumped back to the electrolysis cell to go round the cycle again.

cholesterol: $C_{27}H_{46}O$, the most common animal sterol (steroid alcohol). It can be isolated as white pearly granules which melt at 149 °C. It is only sparingly soluble in water but soluble in oils, fats and solutions of the bile salts. Cholesterol as a chemical has been used as an emulsifying agent in cosmetics and pharmaceuticals, and the main commercial source has been from the spinal cord of beef cattle.

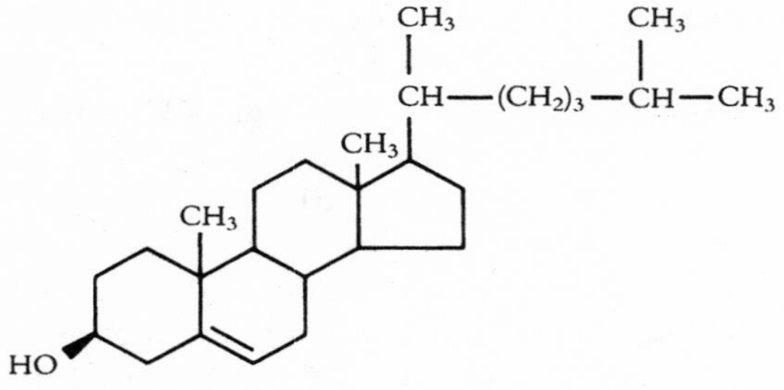

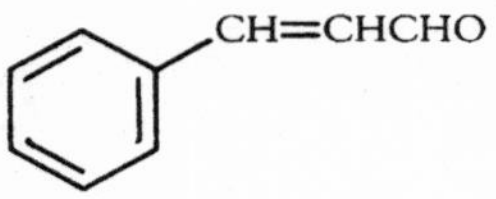

cinnamic aldehyde: C_9H_8O, also known as cinnamaldehyde. It is a yellow liquid with a powerful odour, which can be synthesised from benzaldehyde (C_6H_5CHO) and acetaldehyde (CH_3CHO). It is the main component of cinnamon bark (*Cionnamomum zeylanicum*) and cassia oils. It is more correctly named 3-phenylpropenal.

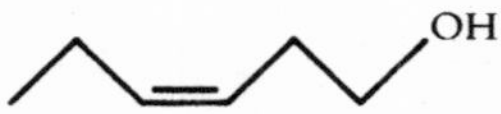

cis-**3-hexenol:** $C_6H_{12}O$, a colourless liquid also known as leaf alcohol which in the pure form has a very powerful, leafy-green smell of newly cut grass. It occurs naturally in various extracts such as geranium, thyme, mulberry leaf, violet and tea.

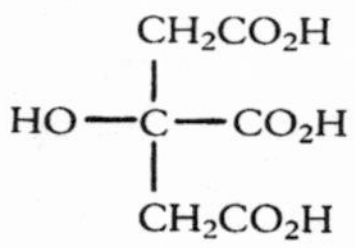

citric acid: $C_6H_8O_7$, more correctly called 2-hydroxy-1,2,3-propanetricarboxylic acid. It is a colourless crystalline material which melts at 153 °C, dissolves readily in water, and has a strongly acid taste. It is present in citrus fruits and can be produced by the fermentation of carbohydrates. Citric acid is used as a flavouring and a food additive (it acts as an antioxidant), and is the acid component of many soft drinks and effervescent tablets. It is also sold as a descaler to remove 'lime' (**calcium carbonate**) deposits. It is non-toxic, and is part of all living cells.

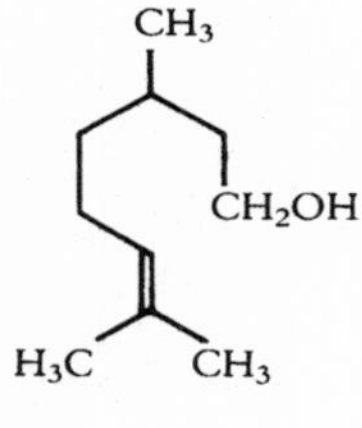

citronellol: $C_{10}H_{20}O$, a colourless liquid that can be obtained by the distillation of geranium or citronella oils. It occurs naturally in these plants and in roses. It is more correctly called 3,7,7-trimethyl-6-octen-1-ol.

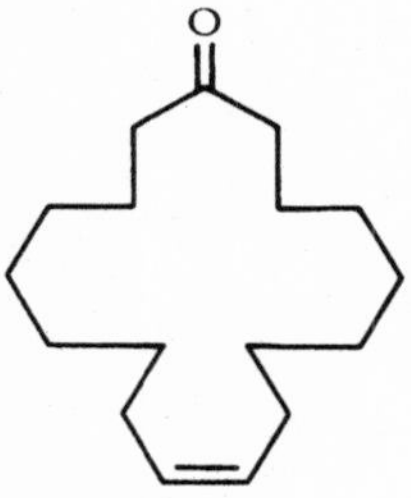

civetone: $C_{17}H_{30}O$, consists of a ring of 17 carbon atoms with an oxygen atom attached to one of them, and a double bond between two of them. It is the main odour chemical of civet, produced by the civet cat. Civetone is able to mould itself to different shapes and this probably enables it to fit different types of odour receptors, hence its tantalising and evocative smell.

cocaine: $C_{17}H_{21}NO_4$, a white crystalline powder which melts at 98 °C and is slightly soluble in water. It is extracted from the leaves of the *Erythroxylon coca* and was once used as a local anaesthetic in dentistry and in so-called tonic wines. It is still widely used illegally as a drug of abuse.

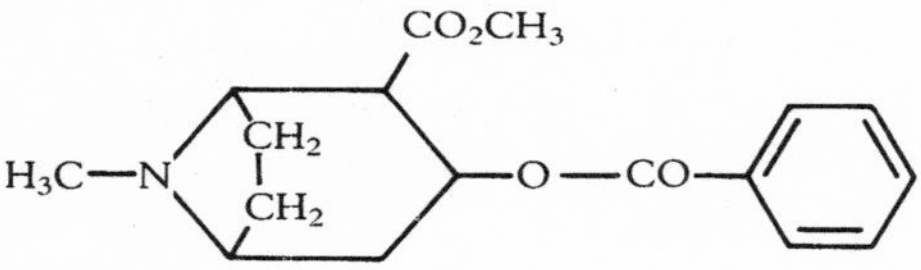

codeine: $C_{18}H_{21}NO_3$, also known as methylmorphine. It is a narcotic alkaloid and is a white solid which melts at 155 °C. Codeine is made from morphine which is extracted from opium, and it is sold as an over-the-counter painkiller, mostly used in conjunction with other painkillers. It is also used in cough syrups.

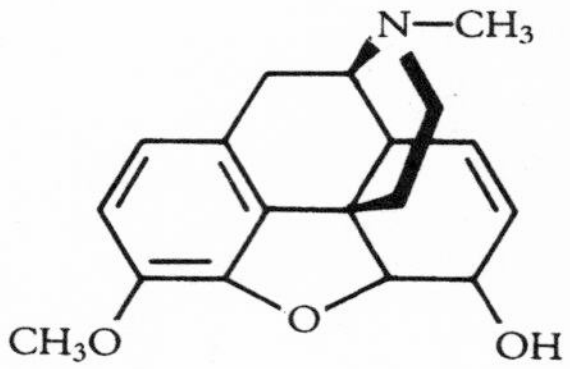

coumarin: $C_9H_6O_2$, a white crystalline powder, melting point 69 °C, also known as benzopyrone. It can be synthesized from salicylic aldehyde and it occurs naturally in many plants and fragrance oils, especially the tonka bean (*Dipteryx odorat*). Coumarin is also found in cassia (Chinese cinnamon tree, *Cinnamomum cassia*) and lavender. It is not permitted in food products because it is a known carcinogen. Coumarin has a sweet, herbaceous, warm, somewhat spicy, odour which when diluted also has the aroma of new mown hay. It is used in deodorising products.

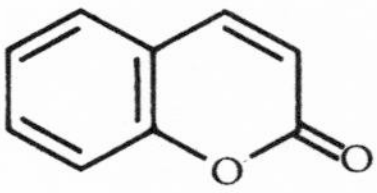

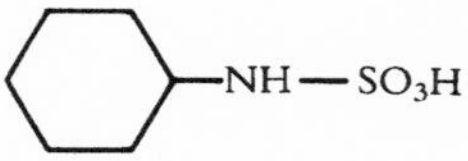

cyclamate: $C_6H_{13}NO_3S$, a white, sweet-tasting crystalline powder once used as an artificial sweetener but now prohibited in some countries.

3-cyclohexenyl-1,3-dioxan: the 'parent' molecule for a new group of synthetic amber-gris odorants. These derivatives vary according to the other chemical groups attached to the molecule, but they all have the same basic chemical structure.

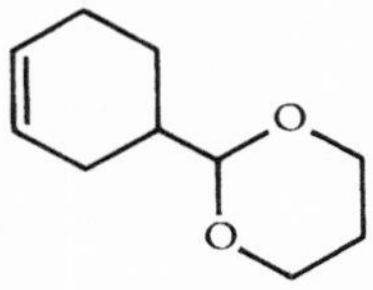

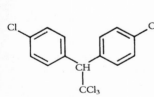

parent

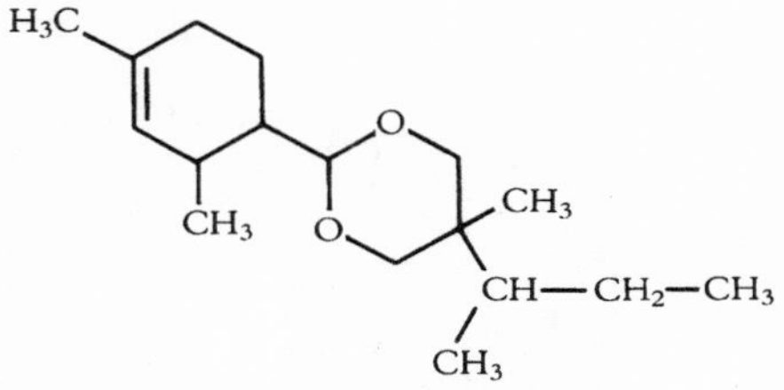

derivative

DDT: short for dichlorodiphenyltrichloro-ethane, $C_{14}H_9Cl_5$, and more correctly called 1,1-bis(4-chlorophenyl)-2,2,2-trichlorothane. DDT comes as colourless crystals which are insoluble in water, and is made from chloral hydrate and chlorobenzene. DDT was first discovered in the 19th century, but its insec-ticide properties were only discovered by accident in 1939. DDT was the first of the new generation of organochlorine insecticides which were much safer than the older variety because they required relatively little to be effective, and they were not toxic to animals. In the 1940s and 1950s they achieved specta-cular success in controlling mosquitos and lifting the threat of malaria from a large part of the globe. Its success led to farmers using it on a wide scale. However, DDT is biode-graded only slowly, and so it built up in the environment and passed along the food chain until it threatened to wipe out some birds of prey. It also gave rise to DDT-resistant insects, and in the 1970s the USA banned its use at home, although not its manufacture for export.

DEHA: this is short for di(2-ethyl-hexyl)adipate which is used as a plasticiser for PVC thereby turning a rigid plastic into one that is supple and suitable for a wide range of applications such as wire insulation, tubing and transparent film.

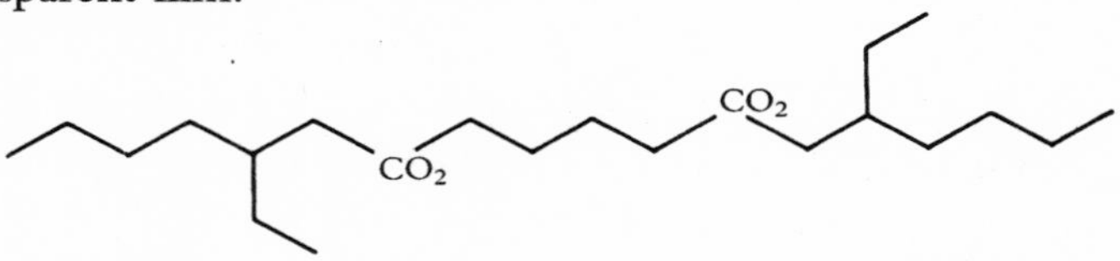

DEHP: this is short for di(2-ethyl-hexyl)phthalate and is an alternative plasticiser for PVC (see **DEHA**).

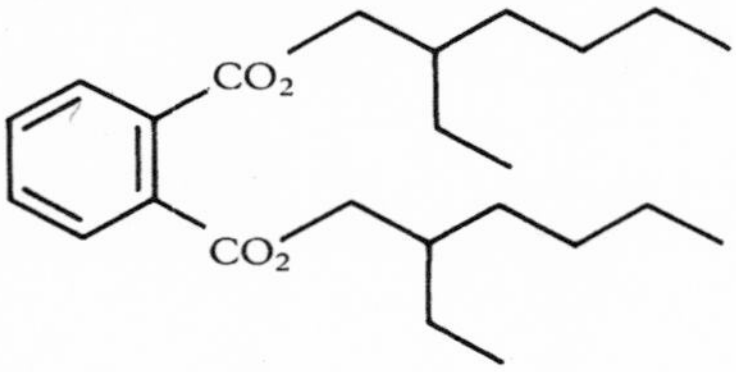

diazepam: $C_{16}H_{13}ClN_2O$, also known as Valium, but its chemical name is 7-chloro-1,3-dihydro-1-methyl-5-phenyl-1,4-benzo-diazepin-2-one. It is a pale yellow powder which melts at 132 °C. It is prescribed as an anti-depressant because it calms the central nervous system, but over-use may lead to addiction. If the chlorine atom in the molecule is replaced by a nitro (NO_2) group then the compound is nitrazepam, better known as the sleeping pill Mogadon.

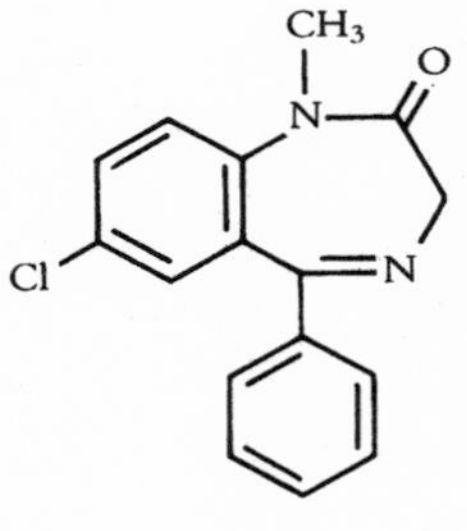

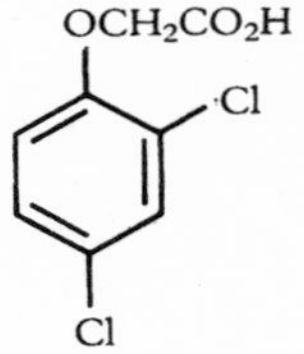

2,4-D: $C_8H_6Cl_2O_3$, the abbreviation for 2,4-dichlorophenoxyacetic acid, where the numbers refer to the positions of the two chlorine atoms attached to the benzene ring. It is a pale yellow/tan solid which melts at 138 °C. It was widely used as a selective weedkiller and also to prevent fruit drop. However, as it was originally manufactured, 2,4-D was contaminated with dioxins. See also **2,4,5-T**.

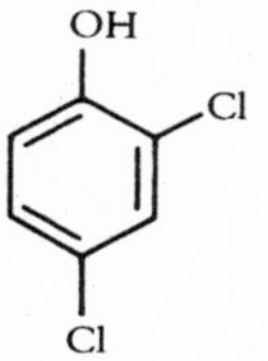

2,4-dichlorophenol: $C_6H_4Cl_2O$. The numbers in the name refer to the positions of the two chlorine atoms attached to the benzene ring. It is a white solid which melts at 45 °C and is produced by the reaction of chlorine gas and phenol, and is used as an intermediate in the manufacture of other compounds.

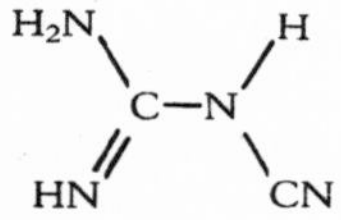

dicyandiamide: $C_2H_4N_4$, used in agriculture to prevent the loss of nitrogen from animal manures. These are susceptible to denitrifying bacteria which end up releasing the nitrogen compounds as soluble nitrates, etc. Dicyandiamide suppresses this microbial action.

diethylene glycol: $C_4H_{10}O_3$, also known as DEG and dihydroxydiethyl ether. This is a colourless, syrupy liquid with a sweetish taste which is used as an antifreeze because when mixed with water it prevents it freezing at temperatures below 0 °C. Diethylene glycol is also used as an industrial intermediate in the production of polymers and resins. It is also used as a plasticiser, a wetting agent, and as a humectant for tobacco (it keeps it moist), cork and sponge.

$$HO—CH_2—CH_2—O—CH_2—CH_2—OH$$

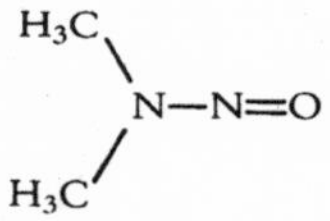

dimethylnitrosamine: $(CH_3)_2NNO$, formed by the chemical reaction of *nitrite* with dimethylamine. As a pure compound it is a dangerous carcinogen. It is also present in minute traces in several foods, especially cooked meats.

dimethyl sulfide: C_2H_6S, also known as methyl sulfide. It is a foul-smelling volatile liquid which boils at 38 °C and is produced naturally by bacteria, along with the gas methyl mercaptan, CH_3SH. Together they are responsible for bad breath. Dimethyl sulfide is sometimes used to 'stench' natural gas so that leaks can be readily detected.

$$CH_3-S-CH_3$$

dioxin: this is the general name for a whole range of compounds whose framework consists of two benzene rings connected via one or two oxygen atoms. Those with two linking oxygens are called dibenzo-dioxins, and those linked through one oxygen are called dibenzo-furans. Each benzene ring can have up to four chlorine atoms, and these are numbered according to the carbon atoms to which they are attached:

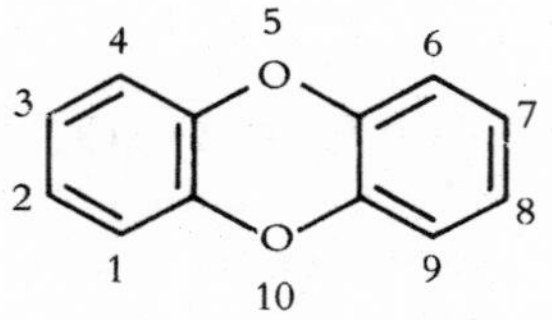

dibenzo-dioxin 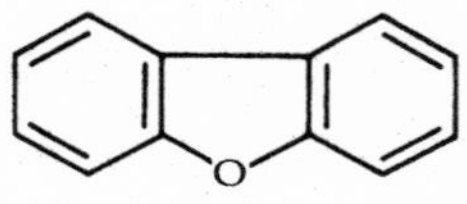   dibenzo-furan

There are many ways of attaching between one and eight atoms to these molecules. For dibenzo dioxin there are 75 possible combinations, and for dibenzo furan there are 135 combinations, making a total of 210 compounds for which the generic term 'dioxins' is used. Of the 75 dibenzo dioxins seven are toxic, and of the 135 dibenzo furans ten are toxic. The most dangerous of all is 2,3,7,8-tetrachlorodibenzodioxin, or 2,3,7,8-TCDD for short, and this is often abbreviated to TCDD. The toxicity of TCDD is the benchmark for all the other dioxins, most of which are only mildly toxic. The key to TCDD's toxicity is the chlorine atoms attached to carbons 2, 3, 7 and 8. Adding one more chlorine atom reduces its toxic action by half. Moving a chlorine from atom 7 to atom 1 reduces it to a thousandth of its original toxicity. When there are the maximum eight chlorine atoms attached to either dibenzo-dioxin or dibenzofuran do we find that the toxic action is negligible. We can see a chemical explanation for this: if 2,3,7,8–TCDD is the key which exactly fits a receptor lock and blocks it, then any changes to the key are bound to make it fit less well.

dolomite: $CaCO_3.MgCO_3$, a naturally occurring mineral that is a mixed calcium magnesium carbonate. It can vary in colour from white to pink to grey. It is from this that magnesium metal is obtained. Dolomite is also used for ceramics, paper making, and as a building material.

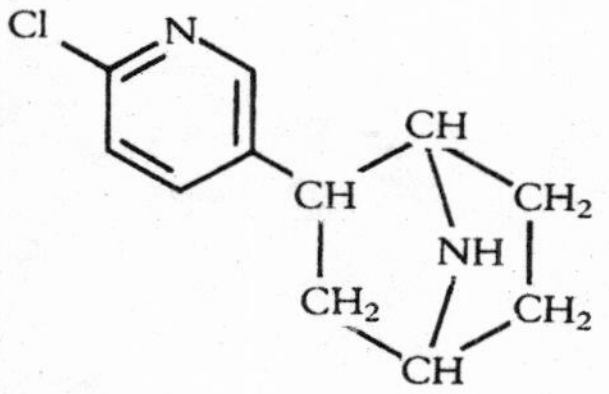

epibatidine: $C_{11}H_{13}N_2Cl$, discovered in 1992 in the skin of the Ecuadorean frog *Epipedobates tricolor*. It is 200 times as potent as a painkiller as morphine.

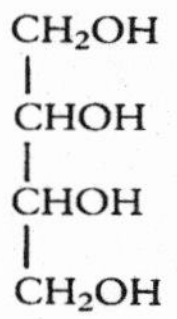

erythritol: $C_4H_{10}O_4$, also known as tetra-hydroxybutane, and more correctly as butane-1,2,3,4-tetrol. This is a white, sweet-tasting, crystalline powder, which melts at 121 °C. It is stable in air, and soluble in both alcohol and water. It is found naturally in *Protocaccus vulgaris* and other lichens.

esters: general formula RCO_2R. This is the name given to the products of the reactions between acids (RCO_2H) and alcohols (ROH) where R stands for the rest of the organic molecule. Esters are generally colourless liquids that have smells characteristic of fruit, and they are often present in fruit.

ethane: C_2H_6, a natural gas which condenses to a liquid at $-89\,°C$. It is the precursor to **ethylene**.

$CH_3—CH_3$

ethanol: C_2H_6O, also known as ethyl alcohol, or just as alcohol. Ethanol is a colourless liquid with a sharp smell. It boils at $78\,°C$ and freezes at $-115\,°C$, which is why it is used in thermometers for measuring low temperatures. It is produced on a large scale, not only by brewing and fermentation but also from **ethylene**, and it is used industrially as a solvent, a fuel, and in the manufacture of paints, plastics, etc. The density of pure ethanol is 0.789 grams per ml, and the density of mixtures of alcohol and water, referred to as the specific gravity, is used as a quick measure of the alcohol content of a liquid.

CH_3CH_2OH

ethyl chloride: C_2H_5Cl, more correctly called chloroethane. This is normally a gas which condenses to a liquid at $12\,°C$, and is used to manufacture tetraethyl–lead, which is the additive of leaded petrol. It has other industrial uses, and once had commercial outlets as an anaesthetic and refrigerant, but along with other chlorine-containing gases it is considered to be potentially damaging to the atmosphere.

$CH_3—CH_2—Cl$

ethylene: C_2H_4, more correctly called ethene, but this name is little used other than in textbooks. Ethylene is a colourless gas (it condenses at $-104\,°C$) which is produced from oil on a vast scale industrially, and is used as the starting point for hundreds of products such as paints, solvents, antifreeze and especially polymers. Linking the two carbon atoms of ethylene is a double bond, and this is the key to its ability to polymerise when heated under pressure, and in the presence of a catalyst:

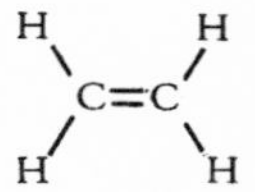

$CH_2=CH_2$ $\xrightarrow{\text{polymerise}}$ $-CH_2-CH_2-CH_2-CH_2-$etc.
ethylene   polyethylene
monomer   polymer

Polyethylene comes in two types, low density polyethylene (LDPE) and high density polyethylene (HDPE), which differ in the length of their chains and in the ways in which they pack together. The conditions of manufacture determine which form is produced. LDPE is used for transparent film and carrier bags, whereas HDPE is used to make plastic containers. Ethylene is produced from natural gas and by ' cracking' oil.

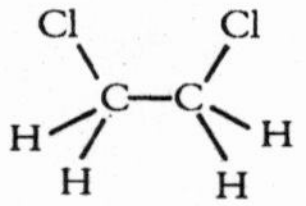

ethylene dichloride: $C_2H_4Cl_2$, has a variety of names including the chemically correct one of 1,2-dichloroethane. It is a colourless, oily liquid which boils at 84 °C, and is the precursor of **vinyl chloride**, which in turn is the precursor of **PVC**. Ethylene dichloride has found several other uses: in leaded petrol, as a paint remover, as a metal degreaser, and as a penetrating agent.

ethylene glycol: $C_2H_6O_2$, also known as glycol, but more correctly called 1,2-ethanediol. This is a clear, colourless, syrupy liquid, which lowers the freezing point of water and is used as an antifreeze. There are many other industrial uses, the principal one being as a solvent for such things as lacquers, resins, printing inks, and adhesives. Despite its wide application, ethylene glycol is toxic, and as little as 100 ccs can kill if consumed.

ethyl heptanoate: $C_9H_{18}O_2$, the ethyl ester of heptanoic acid. It can easily be prepared from ethanol and heptanoic acid. It has a brandy-like aroma.

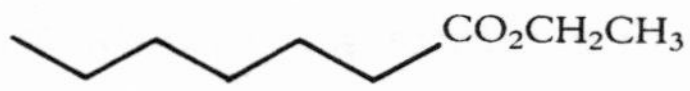

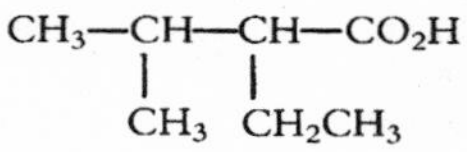

2-ethyl-3-methylbutanoic acid: $C_7H_{14}O_2$, the strong-smelling molecule which gives rum its characteristic aroma.

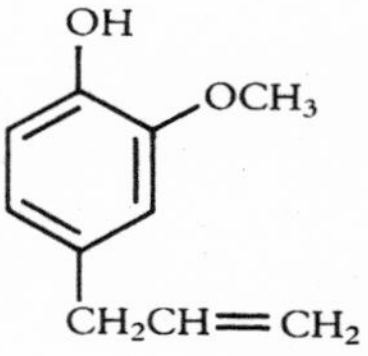

$$CH_3-CH-CH-CO_2H$$
$$\underset{CH_3}{|} \quad \underset{CH_2CH_3}{|}$$

eugenol: $C_{10}H_{12}O_2$, a colourless or pale yellow liquid which boils at 255 °C. It is more correctly called 2-methoxy-4-(2-propenyl) phenol. Eugenol occurs naturally in cloves and is obtained by distilling clove leaf or clove stem oil which can be 80% eugenol. It is also present in cinnamon, pimento and bay leaf. Eugenol has a warm-spicy odour with a slight medicinal hint.

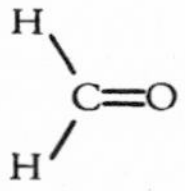

formaldehyde: CH_2O, also known as oxymethylene and more correctly called methanal. It is a gas with a pungent odour which condenses at -19 °C when it rapidly polymerises by reacting with itself. It is generally available as formalin, a 37% solution in water with up to 15% methanol added to prevent polymerisation. This is used as a preservative for biological and medical specimens, and as an embalming fluid. Formaldehyde is produced industrially on a large scale and used to make urea-formaldehyde resins (e.g. melamine), dyes and drugs.

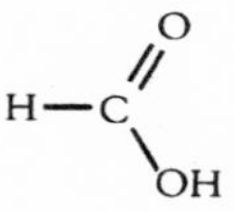

formic acid: H_2CO_2 or HCO_2H, also more correctly called methanoic acid. It is a colourless liquid with a penetrating odour, which boils at 101 °C. It gets its name from *formica*, the Latin word for ant, and it is the main component of the irritant fluid emitted by certain types of stinging ant. Formic acid is produced industrially and used in dyeing, textile-finishing, leather treatment, and the manufacture of fumigants, insecticides, lacquers and solvents. Formic acid is used in electroplating and for silvering mirrors, when it reduces silver nitrate solution to silver which deposits out on the glass surface.

free radicals: these are chemicals that have an unattached electron, and this makes them extremely reactive since the whole object of chemical bond formation is to pair off electrons. Often when a free radical reacts with another molecule it produces yet another free radical among the reaction products. Living systems have to protect themselves against free radicals, and they do this by mopping them up with so-called free radical 'sinks', which are chemicals that will react with a free radical and destroy it, without producing another free radical in its place. We need such defences because the oxygen gas we breathe is prone to reacting to form free radicals.

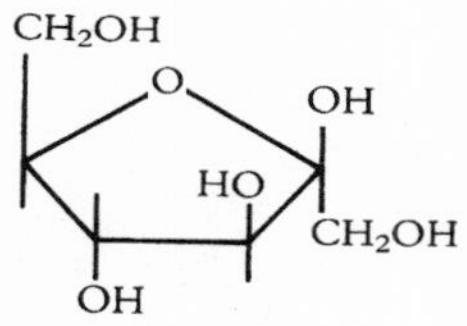

fructose: $C_6H_{12}O_6$, also known as levulose and fruit sugar. It occurs in several fruits and in honey, and is the sweetest of all the natural saccharides. It is a component of nectar. Fructose can be separated as white crystals which melt at 103 °C and decompose, and is obtained commercially from beet sugar or cornstarch by enzyme action. Fructose is the source of energy which powers the movement of sperm.

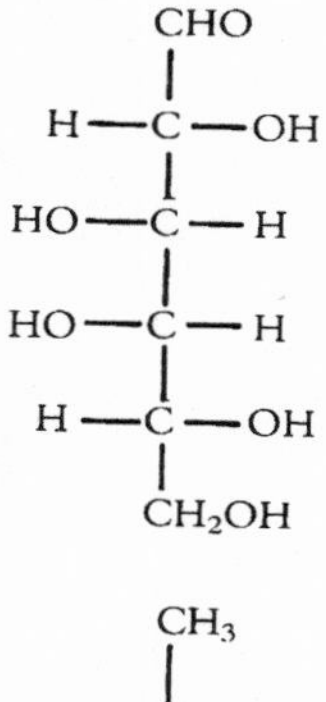

galactose: $C_6H_{12}O_6$, a monosaccharide that is common in milk and other polymeric saccharides. It can be produced from lactose. Galactose is sweet to taste, with a relative sweetness about half that of sugar. It is readily absorbed by the body where it changes to glucose and so raises blood sugar levels.

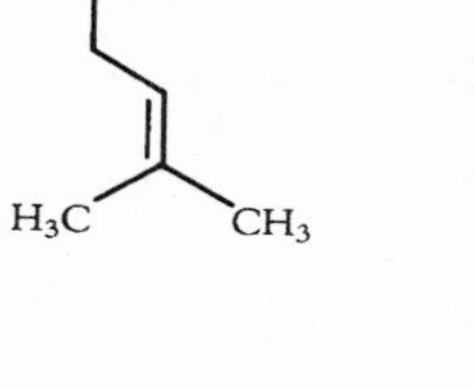

geraniol: $C_{10}H_{18}O$, a colourless liquid which can be made synthetically from pinene or isolated from natural oils of Javanese citronella grass. It is found in geranium, and as geranyl acetate in Eucalyptus oil. Geraniol has a sweet, floral-rose odour with dry undertones.

gluconic acid: **$C_6H_{12}O_7$**, is widely distributed in both animals and plants, usually occurring as part of a larger molecule, as in various gums. It is derived from gum acacia, and separates as needle-like crystals which melt at 165 °C. It is soluble in water and alcohol.

CO₂H
H—C—OH
HO—C—H
H—C—OH
H—C—OH
CH₂OH

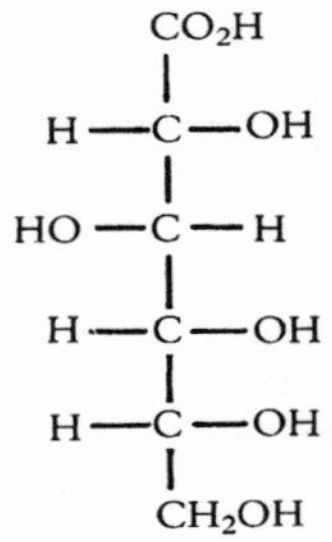

glucose: **$C_6H_{12}O_6$**, more commonly called blood sugar, starch sugar, grape sugar or corn sugar, depending on its source. It exists in two main forms, α-D-glucose and β-D-glucose, of which the former is stable below 50 °C. There is a third form which is a five-membered ring. Glucose occurs in ripe fruits. In the body it represents immediately available energy and circulates in the blood to where it is needed.

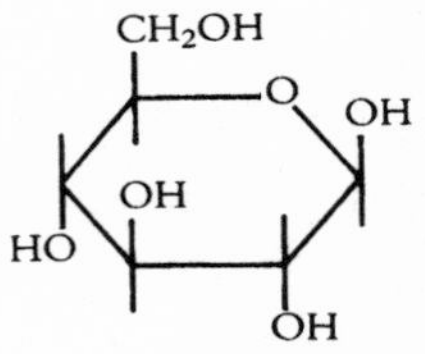

CH₂OH
O
OH
OH
HO
OH
α-D-glucose

glutathione: **$C_{10}H_{17}N_3O_6S$**, γ-L-glutamyl-L-cysteinyl-glycine, a peptide composed of the **amino acids** glutamic acid, cysteine and glycine. It is a component of the living cell and acts as a co-enzyme; it is also involved in transporting amino acids, and in redox-type reactions. Glutathione can be separated as a white crystalline powder which melts at 190 °C and has a mild sour taste. It is used in nutritional and biochemical research.

H₂N
CH
CO₂H
CH₂
CH₂
C
O
NH
CH
CH₂SH
C
O
NH
CH₂
CO₂H

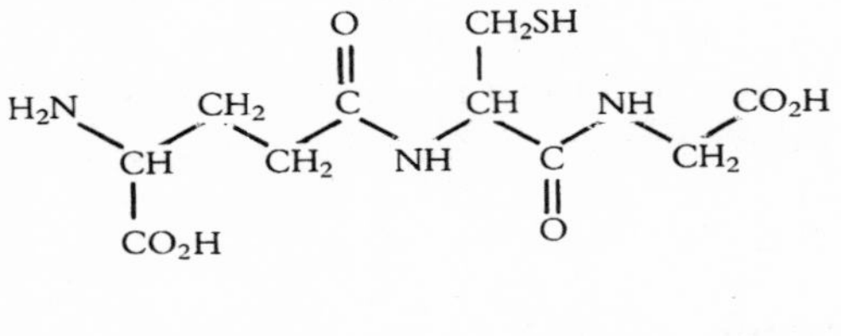

glycerol: **$C_3H_8O_3$**, also known as glycerin or glycerine, but more correctly called propane-1,2,3-triol. It is a clear, colourless, oily liquid which will solidify at 18 °C, boil at 290 °C, and is soluble in water and alcohol. Glycerol is a by-product of soap manufacture

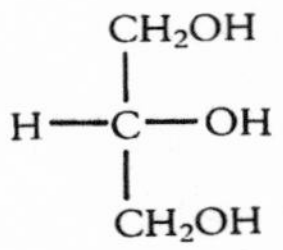

CH₂OH
H—C—OH
CH₂OH

when oils and fats (**triglycer–ides**) are heated with alkalis to give long chain fatty acids and glycerol. It can also be manufactured chemically. Glycerol is used to make many products, among which are resins, pharmaceuticals, cosmetics, emulsifiers, inks, lubricants and antifreeze mixtures.

helium: element, atomic number 2, atomic mass 4, symbol He. This element is completely unreactive and as such exists in Nature only as single atoms. It condenses to a liquid at $-269\,°C$, and behaves like no other fluid in that it can flow up the walls of a container. Helium is part of the atmosphere and is formed in the Earth by radioactive decay—an alpha particle is in effect the nucleus of a helium atom. Helium constitutes 5.2 ppm of the atmosphere, totalling 3.7 billion tonnes (3.7×10^9), but it is slowly escaping away into space. The helium that is used in deep-sea diving, weather balloons and low temperature research instruments comes from natural gas wells, some of which may contain up to 7% helium. World production of helium is about 4500 tonnes a year.

heroin: $C_{21}H_{23}NO_5$, also called diamorphine, or more correctly diacetylmorphine. It is a white, bitter, crystalline compound which melts at $173\,°C$, and is obtained from morphine. Heroin is an illegal drug of abuse which is quickly habit forming and can be fatal in small doses.

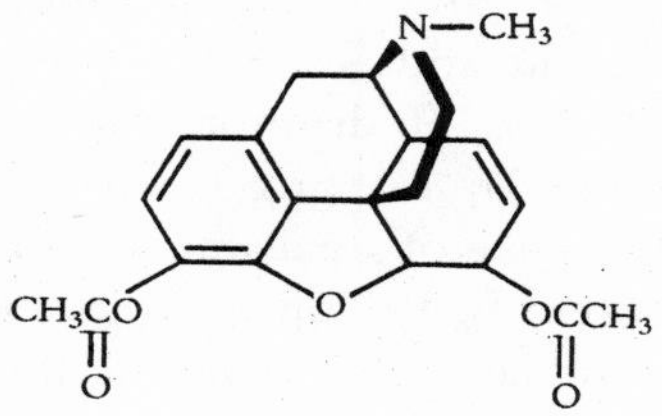

HFCs: this is short for hydrofluorocarbons. These are like the **CFCs**, but they contain no chlorine; only hydrogen and fluorine atoms are attached to the carbon atoms in these molecules. HFCs are replacing CFCs in such uses as refrigerants because even if they eventually escape into the atmosphere they will not affect the ozone layer. Difluoromethane, CH_2F_2, which is being manufactured for use in refrigerators, has a boiling point of $-52\,°C$.

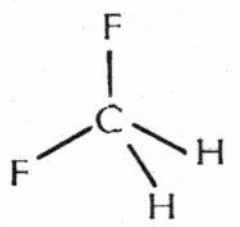

hydrocinnamaldehyde: $C_9H_{10}O$, also known as 3-phenylpropanal. This is a colourless liquid with the odour of hyacinths.

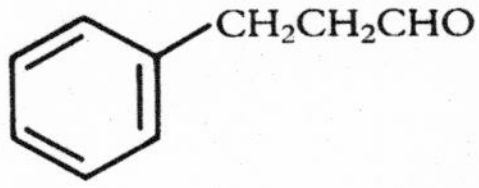

hydrogen: element, atomic number 1, atomic mass 1, symbol H, which exists as a gas of formula H_2. It is colourless, odourless, burns in air, and forms explosive mixtures with oxygen. It is produced on a large scale industrially to the extent of 350 million cubic metres per year, from the controlled reaction of natural gas and oxygen. It is also produced as a by-product of caustic soda manufacture. H_2 gas condenses to a liquid at $-253\,°C$, and can be stored and transported as such. It is used as a rocket fuel and has even been tried as a fuel for buses and cars. Some scientists believe that when natural gas and oil supplies are exhausted, a 'hydrogen economy' will be developed instead.

hydromorphone: $C_{17}H_{19}NO_3$, also known as dimorphone and Novolaudon. It is a narcotic alkaloid and is a white solid which melts at 266 °C. Hydromorphone is made from morphine by reacting it with hydrogen gas in the presence of a palladium catalyst. It is the most potent of the opiate painkillers, and like morphine and diamorphine (heroin) it may be addictive.

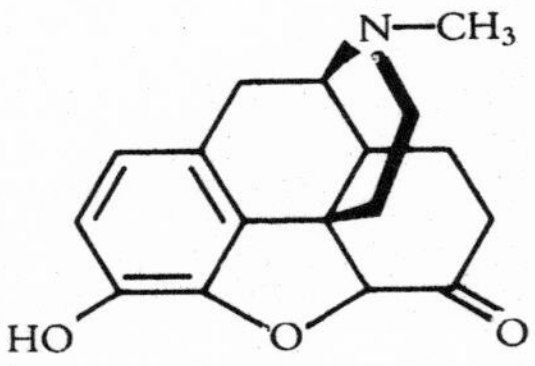

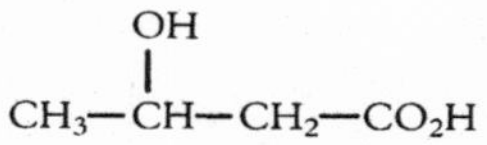

OH
|
CH₃—CH—CH₂—CO₂H

3-hydroxybutyric acid: $C_4H_8O_3$. This is the parent acid from which **polyhydroxybutyrate** polymer (**PHB**) is derived.

ibuprofen: $C_{13}H_{18}O_2$, more correctly called 4–isobutyl–α–methylphenylacetic acid. It is a white powder which melts at 51 °C. it is one of the best-selling over-the-counter pain-killers.

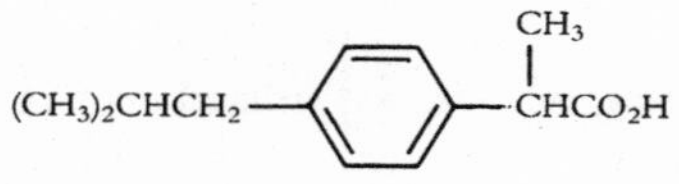

CH₃
|
(CH₃)₂CHCH₂———〈 〉———CHCO₂H

invert sugar: derived from ordinary sugar by the action of acid or enzymes. It gets its name from the effect it has on a beam of plane-polarised light. Ordinary sugar solution will rotate such a beam of light clockwise, a process which chemists call *dextro* (*d*), meaning to the right. When sugar has reacted with water it rotates the beam in the opposite direction, and it is called *laevo* (*l*), meaning to the left. Put another way, it has 'inverted' the light, and so we have the old name of invert sugar.

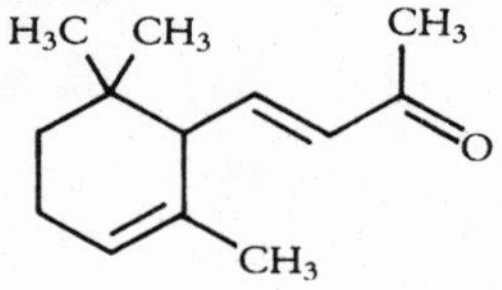

α-ionone

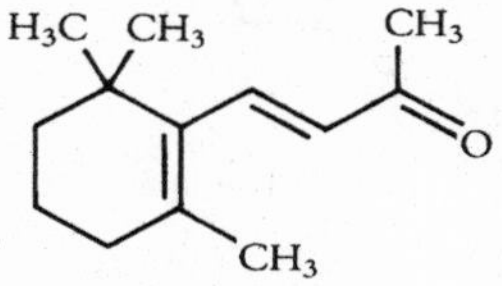

β-ionone

ionone: $C_{13}H_{20}O$, a mixture of two forms, α-ionone and β-ionone. These can be synthesised from citral and acetone. Ionone occurs naturally in *boronia megastigma* which grows wild in Western Australia, and in the oil of violets (*viola odorata*). Synthetic ionone is a pale yellow liquid first made in 1893, and used ever since as a fragrance chemical. α-ionone has a sweet, violet odour, while β-ionone has a green-woody fruity odour reminiscent of cedarwood and raspberries. It also smells of freesias when diluted.

iron: element, atomic number 26, atomic mass 56, symbol Fe (from the Latin *ferrum*). It is a lustrous, workable metal, but rusts readily in damp air. This can be controlled by adding a small amount of carbon (up to 1.7%), and thereby turning it into steel. By adding a lot of nickel and chromium, and traces of other metals, it is possible to turn iron into stainless steel. Iron is not found as the metal itself on Earth (although it is found as such in meteorites). It occurs as the ions, Fe^{2+} and Fe^{3+}, in which the iron atom has lost two or three of its negatively charged electrons. These are called the reduced (Fe^{2+}) and oxidised (Fe^{3+}) forms of iron, and it is as these ions that iron is incorporated into all living things. The average human contains about 4 grams of iron, most being in the blood because this is the element which is at the heart of haemoglobin. Rich sources of iron in the diet are liver and kidney, sunflower seeds, soyabeans, wheat germ, bran and red wine. Iron is the fourth most abundant element in the Earth's crust, and there are vast deposits of haematite (Fe^{3+} oxide), magnesite (Fe^{2+}/Fe^{3+}, oxide), and siderite (iron carbonate).

jasmone: $C_{11}H_{16}O$, which exists in two forms, *cis*-jasmone and *trans*-jasmone. Only the former occurs naturally in jasmin extracts, whereas the synthetic variety contains both. This means that if *trans*-jasmone is detected in a sample of the fragrance, then the material is not solely of plant origin. Jasmone is a pale yellow liquid with a fruity, celery-like odour that changes on dilution to a sweet-floral odour reminiscent of jasmin and cherry blossom. There are several versions of methyl *cis*-jasmonate, because in addition to the *cis* and *trans* forms there can also be variants in which the ester group is on the same side of the five-membered ring as the long chain, or on the opposite side. (There are also mirror-image molecules.) In addition to jasmone there are the methyl jasmonates, also used as fragrances.

cis-jasmone

trans-jasmone

methyl *cis*-jasmonate

krypton: element, atomic number 38, atomic mass 84, symbol Kr. This element is unreactive and as such exists in the air only as a gas composed of single atoms. It condenses to a liquid at $-152\,°C$. Krypton is a trace component of the atmosphere (1.14 ppm) and a small amount is reclaimed from liquid air plants for use in research. An orange–red line in the atomic spectrum of krypton is used as the fundamental standard of length to define the metre; 1 metre is exactly 1 650 763.73 times the wavelength of this line.

lead acetate: $(CH_3CO_2)_2Pb$, once known as sugar of lead. It is one of the few lead salts which is soluble in water, to which it imparts a sweet taste. It is formed by the chemical reaction of acetic acid on lead and was a common, but unsuspected, additive in many alcoholic drinks in former times.

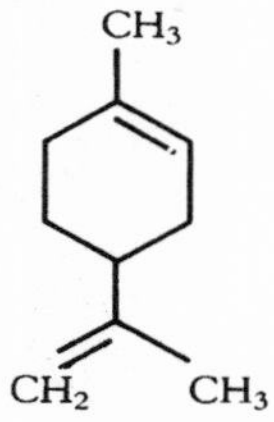

limonene: $C_{10}H_{16}$, a terpene hydrocarbon found in many essence oils from which it can be produced by distillation. It is a colourless liquid with a bright, fresh, clean odour, typical of citrus fruits. It occurs in two mirror–image forms, D-limonene and L-limonene, of which the former predominates.

linoleic acid: $C_{18}H_{32}O_2$, more correctly called 9,12–octadecadienoic acid. It is a doubly unsaturated fatty acid, and is an essential fatty acid in the diet. It is a colourless liquid which freezes at $-12\,°C$ and decomposes on heating to boiling.

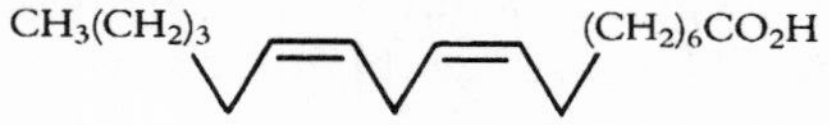

γ**-linolenic acid:** $C_{18}H_{30}O_2$, more correctly called 6,9,12–octadecatrienoic acid. It is a polyunsaturated fatty acid which occurs as the glyceride in many seed fats, and is an essential fatty acid in the diet. Linolenic acid is a colourless liquid which freezes at $-11\,°C$ and boils at $230\,°C$.

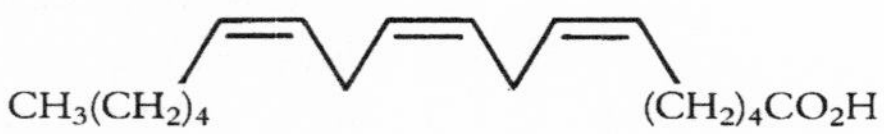

$$CH_3(CH_2)_4 \qquad\qquad (CH_2)_4CO_2H$$

lovastatin: a white crystalline material which melts at $175\,°C$. It is given as a treatment to control high cholesterol levels in the body.

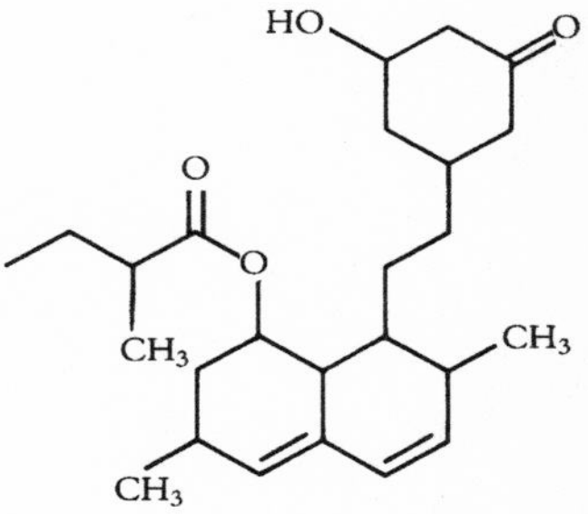

mercury: element, atomic number 80, atomic mass 201, symbol Hg (from the Latin *hydrargyrum*, meaning liquid silver). This liquid metal has been known since ancient times, and is still in production because of its use in industry, street lighting, pesticides and electrical apparatus. Mercury salts are poisonous, and methyl mercury, which can be produced by micro-organisms, is particularly dangerous. This compound is volatile and explains why mercury is widely diffused throughout the environment. Every mouthful of food we eat contains a minute amount of mercury, and some fish such as tuna concentrate it naturally. Humans take in about 5 micrograms of mercury daily, but even if we retained all this within our body we would not accumulate a fatal dose during our lifetime—provided it was in the inorganic form.

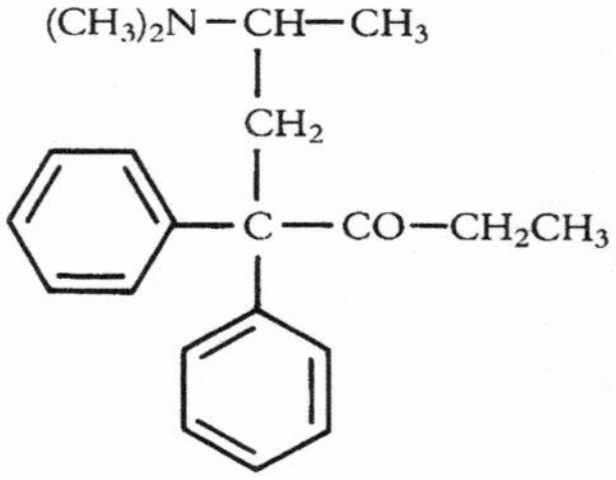

methadone: $C_{21}H_{27}NO$, more correctly called 6-dimethylamino-4,4-diphenyl-3-heptanone. This is a synthetic narcotic, usually supplied as its hydrochloride salt. It is a white solid with a bitter taste which melts at $232\,°C$. Methadone is used in the treatment of heroin addiction, and whereas heroin addiction begins within five days, methadone takes about a month. It can therefore be prescribed to help control the withdrawal symptoms of heroin without itself causing addiction. Methadone can exist in two mirror-image forms, but only the left-hand one is physiologically active.

methane: CH$_4$. This was once called marsh gas because it is the gas given off when organic matter rots under anaerobic conditions. It is the main component of natural gas. Methane is odourless and dangerous, because it can explode in air if sparked. Vast deposits of methane are found in the Earth's crust, and are tapped as an industrial and fuel resource. Industry uses methane mainly as a source of hydrogen, which can be generated by reacting the methane with a limited amount of oxygen. This produces so-called synthesis gas, a mixture of **carbon monoxide** and **hydrogen** which is used to make **methanol**. Methane is formed naturally by certain bacteria living in cows, which explains why these animals and other ruminants contribute to global warming since methane is an atmospheric greenhouse gas.

methanol: CH$_4$O, which is more informatively written as CH$_3$OH to emphasise that one of the hydrogen atoms is attached to the oxygen as an hydroxy (OH) group. Originally methanol was known as wood alcohol because it is given off when wood is heated, and this was how it was once made. Today it is manufactured on a large scale from synthesis gas, a mixture of **hydrogen** and **carbon monoxide** gases. Methanol is used as a solvent, and in some states of the USA as a replacement fuel for petrol because it is cleaner burning. Methanol is also used as a chemical in the manufacture of other compounds such as adhesives and protective coatings.

3-methyl-2-hexenoic acid: $C_7H_{12}O_2$, the unpleasant smelling chemical which is produced in the human armpit.

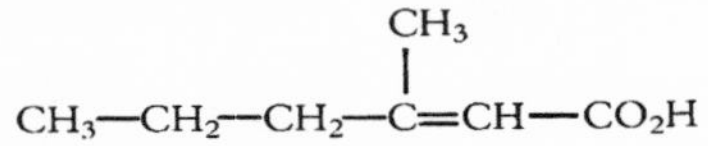

$$CH_3-CH_2-CH_2-\overset{\overset{\displaystyle CH_3}{|}}{C}=CH-CO_2H$$

methyl mercaptan: CH_4S, more informatively written as CH_3SH and more correctly known as methanethiol. This is a gas which condenses to a colourless liquid at 6 °C, and is one of the vilest smelling chemicals known. It is produced as the by-product of bacterial decay of protein under anaerobic conditions, and as such is responsible for bad breath. It is also an industrial chemical that can be made from methanol and hydrogen sulfide, and is an intermediate in the manufacture of methionine (used as a dietary supplement for animals), and in the making of pesticides, catalysts for industry, and the sulfur-based compounds that are used to 'stench' natural gas, so that leaks can be readily detected.

$$CH_3-SH$$

methyl methacrylate: $C_5H_8O_2$, a derivative of ethylene which can be polymerised to form poly(methyl methacrylate), which is also known as Perspex and Lucite. Methyl methacrylate itself is a liquid which boils at 101 °C.

$$CH_2=C\overset{\displaystyle CH_3}{\underset{\underset{\displaystyle O}{\|}}{\diagdown C-OCH_3}}$$

2-methylundecanal: $C_{11}H_{24}O$, also known as aldehyde C12 MNA, short for methyl nonyl acetaldehyde. It is a colourless liquid with a dry, slightly fruity odour reminiscent of ambergris and incense, with floral notes.

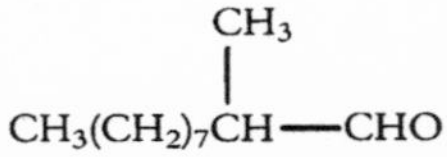

$$CH_3(CH_2)_7\overset{\overset{\displaystyle CH_3}{|}}{CH}-CHO$$

mevalonic acid: $C_6H_{12}O_4$, more correctly called 3,5-dihydroxy-3-methylpenantoic acid. This is a precursor to cholesterol formation in the body. It is an oil which occurs as a mixture of two interchanging molecules.

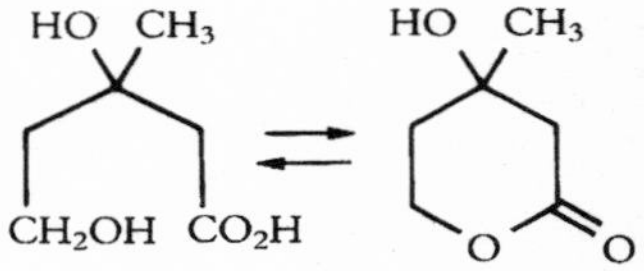

mevastatin: $C_{23}H_{34}O_5$, given as a treatment to control high cholesterol levels in the body.

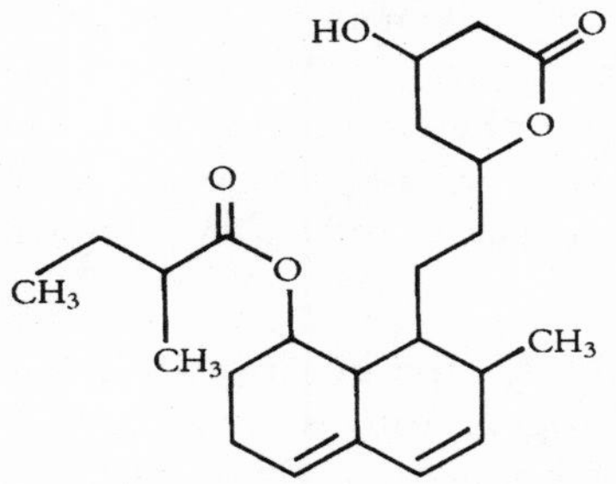

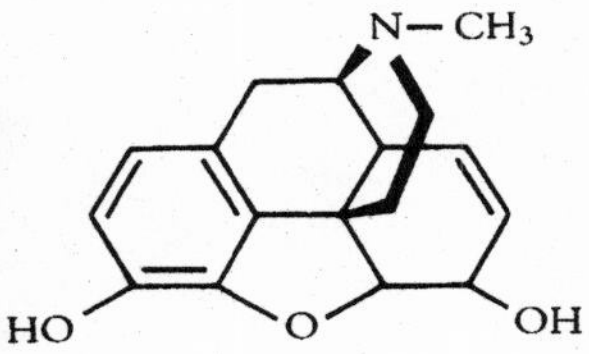

morphine: $C_{17}H_{19}NO_3$, the active constituent of opium, which oozes as a milky juice from unripe poppy seeds of the *Papaver somniferum*. Pure morphine is a white solid, slightly soluble in water. It is among the most powerful painkillers known, but it is addictive.

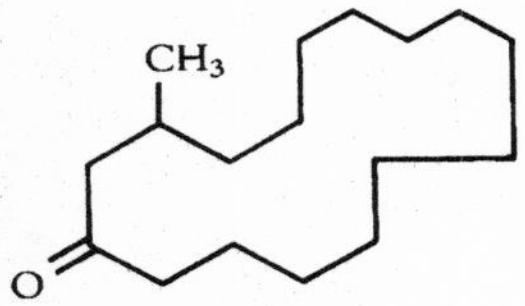

muscone: $C_{16}H_{30}O$, a white or colourless crystalline solid with a very soft, sweet, musky, erogenous odour. It is the main aroma component of natural musk, but is now made synthetically from diacetyldodecane. It is chemically similar to **civetone**.

N-**acetylcysteine:** see acetylcysteine.

neon: element, atomic number 10, atomic mass 20, symbol Ne. This element is completely unreactive and as such exists in the air only as a gas composed of single atoms. It condenses to a liquid at $-246\,°C$. Neon is not very abundant on Earth and constitutes 18 ppm of the atmosphere, although this adds up to a total of 65 billion tonnes (65×10^9). About one tonne is extracted each year from liquid air for use in ornamental lighting (neon signs).

nicotine: $C_{10}H_{14}N_2$, chemical name β-pyridyl-α-*N*-methylpyrrolidine. Nicotine is an alkaloid which is present in tobacco and can be extracted as a viscous colourless oil which turns brown on exposure to air. In the small amounts taken in by smoking tobacco, nicotine has a pleasurable effect on the brain, but it is nevertheless a highly toxic chemical, and is addictive. In the form of its salts, nicotine has been used as an insecticide and fumigant.

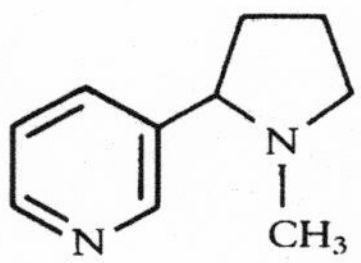

nitrate: NO_3^-, the negative ion in which a central nitrogen atom is surrounded by, and chemically bonded to, three oxygen atoms. Nitrates are salts of nitric acid in which the hydrogen of HNO_3 has been replaced by a positive metal ion, leaving the nitrate as the negative ion, NO_3^-. The best known salts are sodium nitrate, $NaNO_3$, which is found as large deposits in Chile and called Chile nitre or Chile saltpetre, and potassium nitrate, KNO_3, which is the original saltpetre. All nitrates are very soluble in water.

nitric acid: HNO_3, a strong, oxidising acid. In its concentrated form, it gives off brown fumes of **nitrogen dioxide** (NO_2). This acid is very dangerous and will rapidly char any organic material it comes in contact with. Nitric acid is made in large quantities by the oxidation of ammonia to form **nitric oxide**, NO, and then reacting this with air and water to form HNO_3. Most nitric acid is used in the manufacture of **ammonium nitrate**, and the rest is used to make plastics, dyes and explosives.

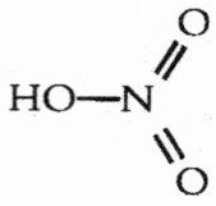

nitrite: NO_2^-, a molecular ion and technically the salt of nitrous acid HNO_2, although this is very unstable. Of the nitrite salts, sodium nitrite, $NaNO_2$, and potassium nitrite, KNO_2, are the most common, and are used in curing meat and as chemical reagents in the laboratory. Nitrite is radically different from nitrate in its chemical behaviour; for example, whereas nitrate is relatively indifferent towards metals, preferring to exist independently as NO_3^-, nitrite is attracted to metals such as iron, and bonds to them very strongly.

nitrogen: element, atomic number 7, atomic mass 14, symbol N. It occurs in the air as N_2, and is more correctly known as dinitrogen. It is a colourless, unreactive gas which comprises 79% of the atmosphere. It is produced on a large scale industrially by liquefying air, and is shipped as liquid nitrogen which boils at $-196\,°C$. It is used in metallurgy, food packaging, refrigeration, and to manufacture ammonia.

nitrogen oxides: most likely to be encountered as pollutants, because they form during combustion and are generally labelled together as NO_x. There are three gases: nitrous oxide, N_2O, was once known as laughing gas, following its use as an anaesthetic in dentistry, when it led to some patients becoming euphoric. It occurs naturally in the atmosphere in low concentration and it is a relatively unreactive gas. Nitric oxide, NO, is produced in the human body which uses it as a chemical messenger. Its best known role is in securing an erection in males, but it has several other roles to play, such as relaxing arteries. Nitric oxide is chemically very reactive, and is easily oxidised by oxygen to nitrogen dioxide. It has a powerful affinity for metals. Nitrogen dioxide, NO_2, is a brown gas with a choking effect if breathed in large amounts due to its instant reaction with water in the lungs to form nitric acid. It is very reactive towards organic molecules, and under atmospheric conditions is a key factor in producing smog because it combines with unburnt hydrocarbon emissions in the exhaust gases of cars.

nylon: now a general name for a family of polymers which are polyamides, meaning they have the molecular amide unit, $-CO-NH-$, along the backbone of the polymer. The best known nylon is nylon-6,6 in which amide units are linked by chains with six carbon atoms. This was the nylon discovered by Wallace Carothers in 1935. Other nylons have carbon chains of different lengths. Nylon is strong and versatile and finds use in tyres, hosiery (for which it is spun so fine that its inherent strength is nullified), bristles, ropes, fishing nets, sails, carpets, clothing, upholstery, membranes, gears and bearings.

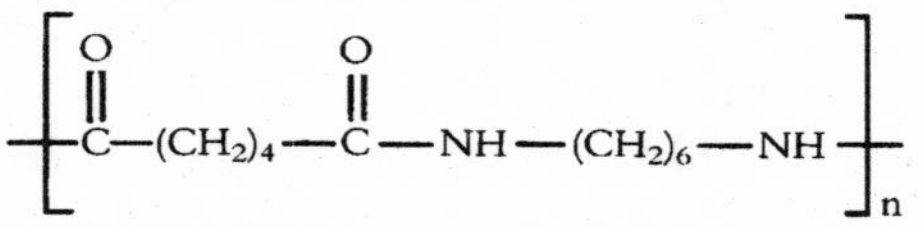

$$\left[\begin{array}{c} \overset{\displaystyle O}{\overset{\|}{C}} - (CH_2)_4 - \overset{\displaystyle O}{\overset{\|}{C}} - NH - (CH_2)_6 - NH \end{array} \right]_n$$

octanol: $C_8H_{18}O$, more correctly called 1-octanol, and also known as *n*-octyl alcohol. It is a colourless liquid which boils at 195 °C. It is used in perfumery, cosmetics and in industry as a solvent and anti-foaming agent. It is also an intermediate for making other chemicals.

$CH_3(CH_2)_6CH_2OH$

Opren: $C_{16}H_{12}ClNO_3$, has the generic name benoxaprofen, and the chemical name 2-(4-chlorophenyl)-α-methyl-5-benzoxazole acetic acid. It is an anti-inflammatory analgesic, which is now banned in some countries.

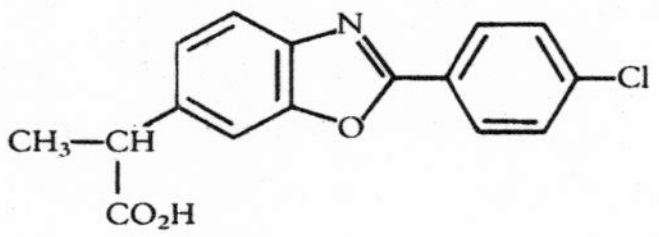

oxygen: element, atomic number 8, atomic mass 16, symbol O, which exists as a gas of formula O_2 and is more correctly called dioxygen. It is colourless, odourless, very reactive, and forms oxides with almost every other element. Despite this it accounts for 21% of the Earth's atmosphere. There are limitless supplies of the gas to be obtained from liquid air and about 100 millions tonnes are extracted from this source every year for use in steel making, metal cutting, in the chemicals industry and for medical applications.

ozone: the other gaseous form of elemental oxygen, with formula O_3. It is an unstable blue gas with a pungent odour that can be detected near electric sparks and UV lamps. It will liquefy at $-12\,°C$ but is highly dangerous as such. The gas is produced in ozonisers which work by passing an electric discharge through air and converting a few per cent of the O_2 to O_3. The gas is then used as an oxidising agent in industry, and as a disinfectant for swimming baths and public water supplies. Ozone is a natural part of the upper atmosphere and serves to filter out dangerous ultraviolet light from the Sun's rays. Ozone is sensitive to chlorine atoms which can destroy it, and hence the concern about long-lived organochlorine vapours in the atmosphere, such as the CFCs. In the lower atmosphere ozone is produced as part of photochemical smog, and as such is damaging to plants and human lungs.

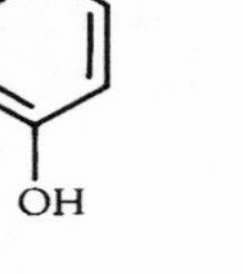

NHCOCH$_3$

OH

paracetamol: $C_8H_9NO_2$, also known as N-acetyl-*para*-aminophenol or N-acetyl-4-aminophenol. It exists as colourless crystals which melt at $168\,°C$. Paracetamol is slightly soluble in water, and has a bitter taste. Although it is best known as a painkiller, it is also used as an intermediate in the production of pharmaceuticals, dyes and photographic chemicals.

PCB: the abbreviation for polychlorinated biphenyl, and refers to a collection of molecules that have the common biphenyl (also known as diphenyl) framework. In the middle years of this century the PCBs were produced on a large scale as a solvent and particularly as an insulating fluid for electrical transformers. Their escape into the environment has given cause for concern because, like dioxins, these chemicals are long-lived and slightly carcinogenic.

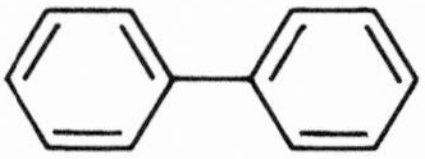

biphenyl

PCDD: the abbreviation for pentachlorodibenzodioxin, in which there are five chlorine atoms attached to the dibenzodioxin molecule (see **dioxin**).

PET: the abbreviation for polyethylene terephthalate, which is the chemical name for the polymer often referred to simply as polyester. PET can be spun into fibres (Terylene, Dacron) which produce fabrics that resist creasing, and are used for clothing. The polymer can also be used to make transparent film and containers, such as bottles.

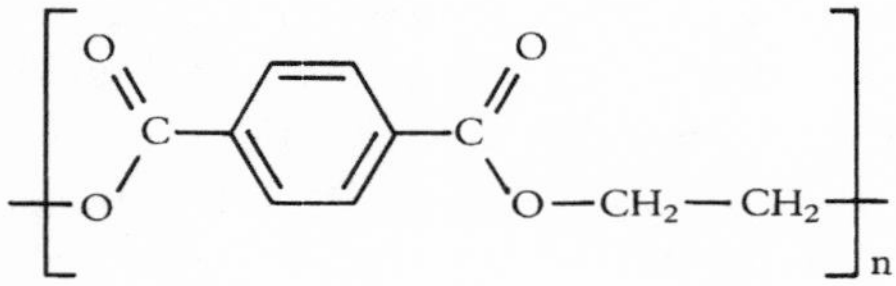

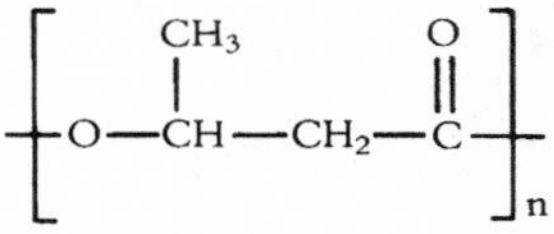

PHB: the abbreviation for polyhydroxybutyric acid, a natural polymer made by a wide range of micro-organisms, such as the bacterium *Alicaligenes eutrophus*, as a convenient way of storing energy. PHB consists of chains of **3-hydroxybutyric acid** which has an acid group CO_2H at one end of the molecule and a hydroxyl group OH at the other. When these react they form a chemical bond called an ester link, and in this way they can join head-to-tail to form long chains thousands of units long, in other words a polymer.

phenol: C_6H_6O, also known as carbolic acid and once commonly used as an antiseptic. It is now considered too dangerous for general use, because it is an irritant to skin. Phenol is a pinkish crystalline solid (although white when pure) which melts at 43 °C and has a distinctive odour. It is still manufactured on a large scale and used to produce products ranging from aspirin to resins. The term phenol also applies to the group of compounds characterised by a hydroxyl group (OH) attached to a benzene ring.

phenylacetic acid: $C_8H_8O_2$, also known as α-toluic acid, and consisting of white crystals which melt at 76 °C. It has a sweet, floral odour and is the main aroma associated with honey, in which it occurs as the ethyl ester. It is made synthetically from benzyl cyanide and hydrochloric acid. It is used as a flavouring, and in the manifacture of penicillin G.

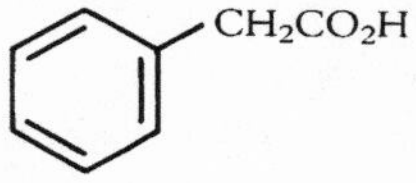

CH₂CO₂H

acid

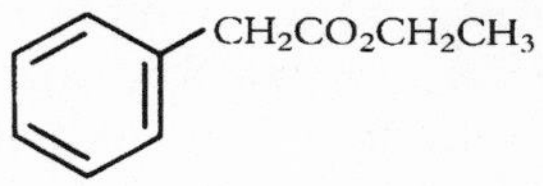

CH₂CO₂CH₂CH₃

ester

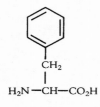

phenylalanine: $C_9H_{11}NO_2$, an essential amino acid also named α-amino-β-phenyl-propionic acid. It is soluble in water, from which it can be obtained as plate-like and leaf-like crystals; it decomposes at 283 °C. Phenylalanine is isolated commercially from proteins and is used in biochemical research and sometimes as a dietary supplement.

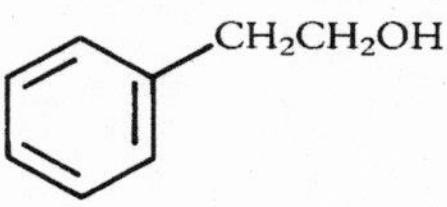

2-phenylethanol: $C_8H_{10}O$, a colourless liquid alcohol which can be synthesized from benzene and ethylene oxide. It occurs naturally in rose, geranium and neroli oils and has a mild,

warrm, rose-like odour with green, hyacinth
undertones.

phenylethylamine $C_8H_{11}N$, also known as
2-phenylethylamine or 1-amino-2-phenyl-
ethane. This is a liquid which boils at
195 °C, and has a fishy odour in the concentrated state. It occurs naturally in chocolate
and a 100 gram bar can contain as much as 0.7
grams of phenylethylamine. This chemical is
believed to raise blood pressure, producing a
heightened sensation and a feeling of well-
being.

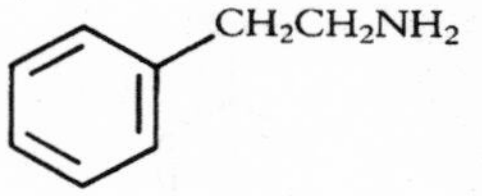

polyethylene: sometimes abbreviated to PE.
This polymer consists of endless chains of
carbon atoms, each carrying two hydrogens
(see **ethylene**).

polyhydroxybutyrate (PHB): a biodegradable polymer sold under the trade name of
Biopol (see **PHB**).

polystyrene: $(C_8H_8)_n$. This is a thermoplastic resin, which means that it softens
when heated but hardens again when it
cools. It is hard and strong, with excellent
electrical resistance. Expanded polystyrene
foam finds use as thermal insulation.
However polystyrene has two drawbacks: it
is attacked by organic solvents, and it needs
additives to protect it against sunlight (see
also **styrene**).

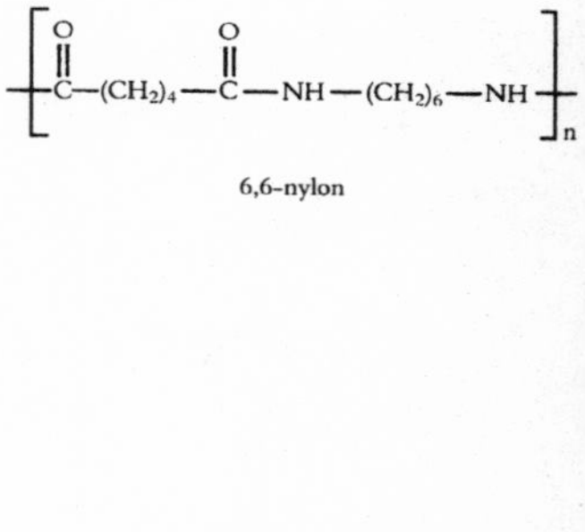

6,6-nylon

polyvinyl chloride: see PVC.

propane: C_3H_8, the third member of the
alkane series after **methane** and **ethane**. It is
also a gas, but with a higher boiling point
(-43 °C) than its lighter fellows. It is used as
an industrial resource, a fuel, a refrigerant and
an aerosol propellant. It is highly flammable.

$CH_3-CH_2-CH_3$

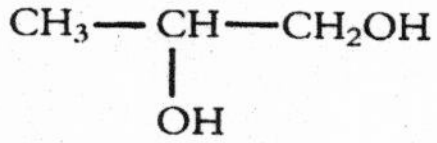

$$CH_3 - CH - CH_2OH$$
$$\quad\quad\ |$$
$$\quad\quad OH$$

$$CH_3 - CH_2 - CH_2OH$$
1-propanol

$$CH_3 - CH - CH_3$$
$$\quad\quad\ |$$
$$\quad\quad OH$$
2-propanol

propane-1,2-diol alginate: this is the ester of alginic acid (see **alginate**). Propane-1,2-diol has the formula shown.

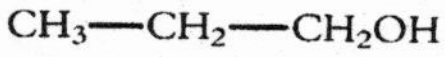

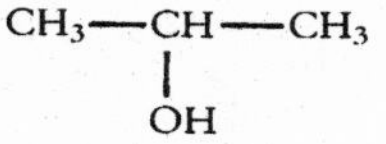

propanol: C_3H_8O, formerly known as *n*-propyl alcohol, but more correctly called 1-propanol. There is also the related chemical 2-propanol. Both alcohols are colourless liquids which boil at 97 °C (1-propanol) and 82 °C (2-propanol), and both are used industrially as intermediates in the manufacture of other chemicals.

prostaglandins: these are a group of physiologically important compounds derived from arachidonic acid, which is a 20-carbon fatty acid that is found in glands and in the liver. Prostaglandins have a role to play in hormones, reproduction, blood pressure, digestion, smooth muscle stimulation and labour. They are particularly abundant in semen. Prostaglandins have been intensively studied and they can be made synthetically, although the most prolific source is from other organisms such as the gorgonian sea whip which is abundant in the Caribbean.

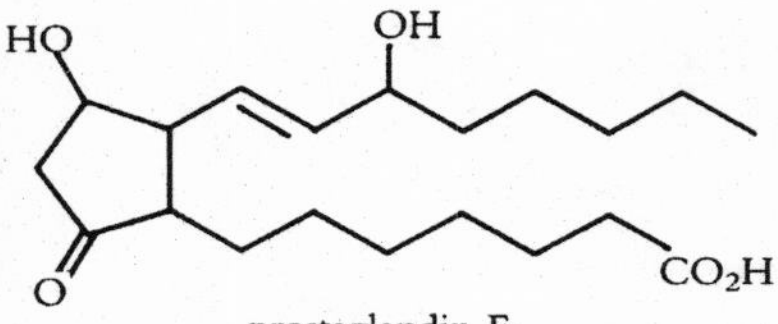

prostaglandin E₁

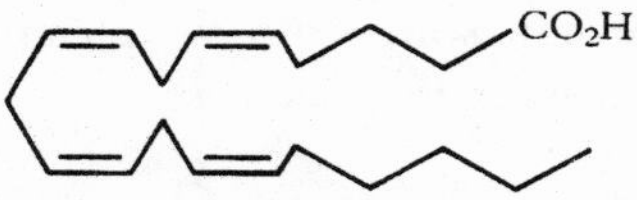

arachidonic acid

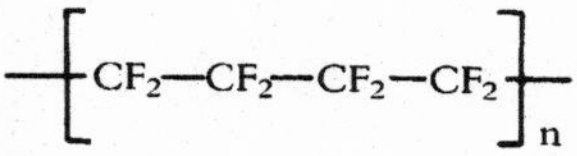

$$\left[CF_2 - CF_2 - CF_2 - CF_2 \right]_n$$

$$\left[\begin{array}{c} CH - CH_2 - CH - CH_2 \\ |\quad\quad\quad\quad | \\ Cl \quad\quad\quad\quad Cl \end{array} \right]_n$$

PTFE: $(C_2F_4)_n$, poly(tetrafluoroethylene). This polymer is known by a variety of trade names such as Teflon. It has a very low coefficient of friction and this makes it an ideal coating for non-stick pans and skis.

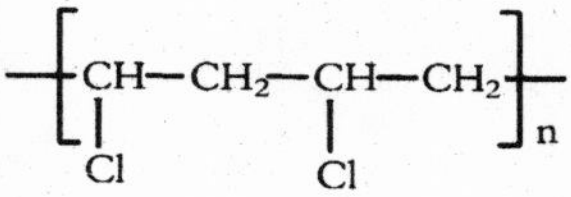

PVC: $(C_2H_3Cl)_n$, poly(vinyl chloride). This polymer is made from vinyl chloride, and is a thermoplastic material. It is an excellent water repellent and resists weathering, so finds many outdoor uses. It is easily coloured and shaped and can be produced as fibre, foam or film.

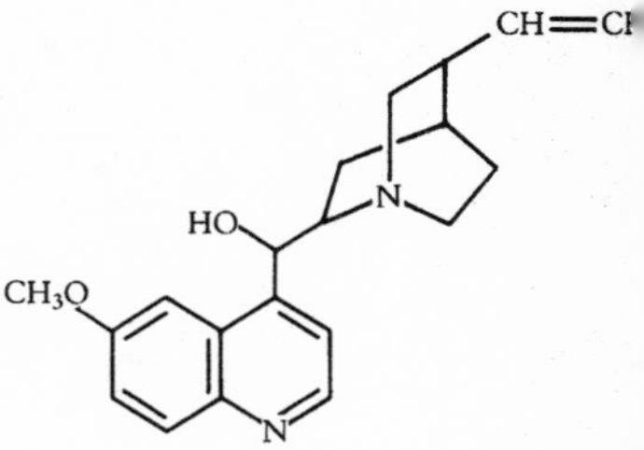

quinine: $C_{20}H_{24}N_2O_2$, an alkaloid extracted from the bark of the *cinchona* tree of South America. It can be obtained as a white powder with a bitter taste. It is slightly soluble in water and is the flavouring used to make tonic water. It was the original antimalarial drug and it operates by inhibiting replication of DNA.

quinoline: C_9H_7N, a colourless liquid that darkens with age and boils at 238 °C. It can be obtained from coal tar. When used in fragrances it is most likely to be as its *iso*butyl derivative, which has an intense leather odour with wood and dry undertones, and is employed in chypre-type perfumes.

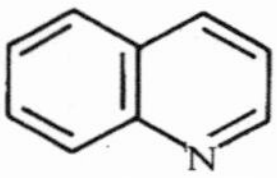

quinoline

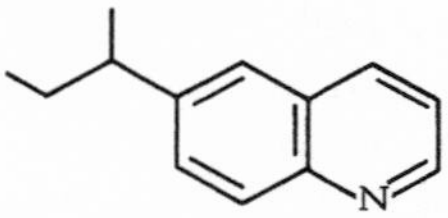

6-*iso*-butylquinoline

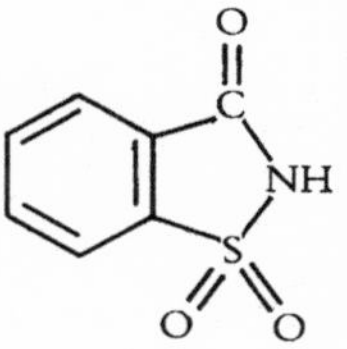

saccharin: $C_7H_5NO_3S$, a white powder, stable in air, soluble in water, and about 300 times sweeter than sucrose.

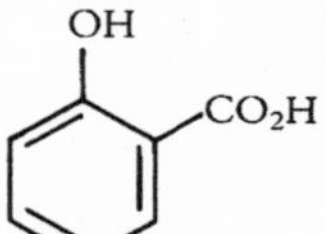

salicylic acid: $C_7H_6O_3$, more correctly called 2-hydroxybenzoic acid. It is a white powder with an acrid taste which melts at 158 °C. Although it can be obtained from natural sources it is made from the sodium salt of phenol by reacting with carbon dioxide. Salicylic acid is an intermediate in the production of aspirin, resins, dyestuffs, and fungicides.

salt: $NaCl$, also known as sea salt, rock salt, table salt and the mineral halite. The chemical name is sodium chloride. Large crystals are colourless, although they look white when small. Salt is an essential dietary component, and it has many general uses. Large deposits are mined, and salt is a major resource for the chemicals industry because it is used to make **chlorine** gas and **caustic soda** (sodium hydroxide).

silicon dioxide: SiO_2, occurs naturally in various forms such as common sand, quartz, flint, and even as semi-precious gemstones such as rhinestones (found in the River Rhine, hence the name). Silicon dioxide is not a simple molecule, but consists of an array of silicon atoms each surrounded by four oxygens which link to other silicons.

sodium alginate: see alginate.

sodium bicarbonate: $NaHCO_3$, a white solid that is also known as baking soda, and more correctly as sodium hydrogen carbonate. It is soluble in water to give an alkaline solution, and is commonly taken as an antacid. Its reaction with acids generates carbon dioxide, and the reaction is also used to make fizzy tablets. $NaHCO_3$ is made by passing carbon dioxide into a solution of sodium carbonate, Na_2CO_3. It is produced on a large scale and most is used in the food industry.

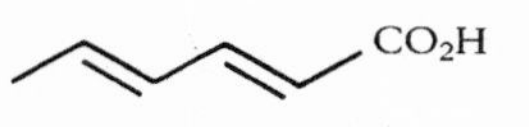

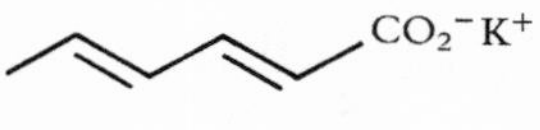

Sorbic acid: $C_6H_8O_2$, more correctly called 2-*trans*-4-*trans*-hexendienoic acid. This is a white crystalline solid which melts at 135 °C and is only slightly soluble in water, although its salts, such as potassium sorbate, are much more soluble. Sorbic acid and sorbates are used as food preservatives since they kill fungi and inhibit moulds.

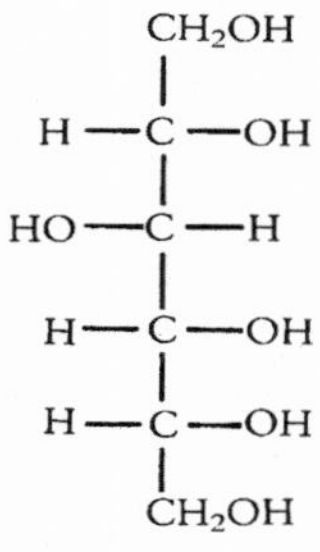

sorbitol: $C_6H_8(OH)_6$, also known as hexahydric alcohol. It is a white crystalline powder with a faint sweet taste, and which absorbs moisture from the air. It is soluble in water and organic solvents, and melts at 95 °C.

Sorbitol is present in some fruits and berries, but is manufactured from glucose by reacting it with hydrogen under pressure. It is used in cosmetic creams, toothpaste, candy, and in industry as a stabiliser for resins, a thickener, an emulsifier, and an anticaking agent.

steroids: this is a group of biologically important chemicals which share the same basic structure of four inter-linked rings. An example is **androstenone**.

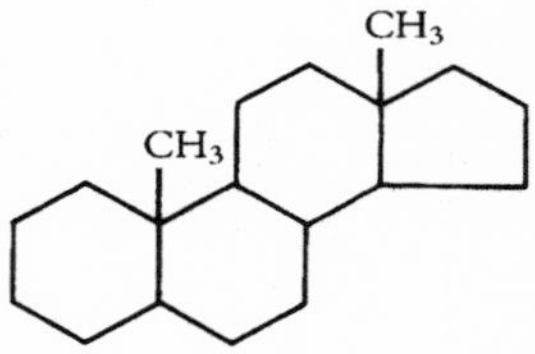

streptomycin: $C_{21}H_{39}N_7O_{12}$, an antibiotic. The term is also used for a group of antibiotics produced by actinomycetes. Streptomycin itself is got from *streptomyces griseus*, and is active against gram–negative bacteria and the tubercule bacillus.

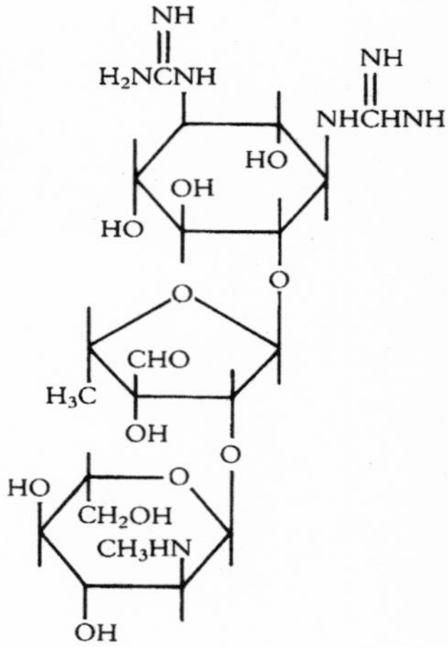

styrene: C_8H_8, more correctly called phenyl-ethene. It consists of ethylene with a benzene ring attached to one of the carbon atoms. It is a colourless, oily liquid which boils at 145 °C, and which readily undergoes polymerisation to **polystyrene**.

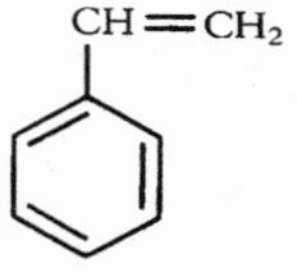

Sucralose: $C_{12}H_{19}Cl_3O_8$, more correctly called 1,6–dichloro–1,6–dideoxy–β–D–fructo-furanosyl 4–chloro–4–deoxy–α–D–galactopyrano-side. Sucralose is the trade name for a type of chlorinated saccharide in which three chlorine atoms have replaced hydroxy groups of **sucrose**.

sucrose: $C_{12}H_{22}O_{11}$, more correctly called saccharose; its full chemical name iso-β-D-fructofuranosyl-α-D-glucopyranoside. It is also known as cane sugar, beet sugar, or just sugar. It melts when heated above 160 °C and caramelises above 200 °C.

sulfanilimide

sulfanilimide: $C_6H_8N_2O_2S$. This is the simplest member of a group of 50 or so drugs which go under the general name of the sulfanilamides or sulfa drugs. They share the common chemical grouping shown in the parent compound.

sulfur dioxide: SO_2 is formed when any sulfur-containing compound burns. It is a colourless gas with a choking odour, and it is soluble in water forming a solution of a weak acid, sulfurous acid, H_2SO_3, which is used as a preservative in food processing, especially for fruit, and to prevent bacterial infection in grape juice prior to fermenting. The SO_2 which escapes to the air from coal-fired power stations is partly responsible for acid rain. Industrially SO_2 is made by burning sulfur, or 'roasting' iron pyrites (FeS). Most SO_2 is used to make sulfuric acid, by first oxidising SO_2 to sulfur trioxide, SO_3, and then dissolving this in water:

$SO_3 + H_2O \rightarrow H_2SO_4$ (sulfuric acid)

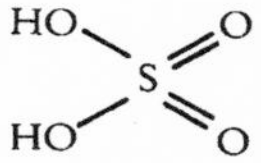

Other uses of SO_2 include paper manufacture, bleaching, and in the production of solvents and detergents.

tannin: this is the general name for a large group of phenol compounds derived from plants. They are bitter tasting and occur in small amounts in alcoholic drinks which have been left to mature in oak casks, from which the tannins are extracted. The tannins are derivatives of tannic acid, also known as gallotannic acid, $C_{76}H_{52}O_{46}$, and this can be used to clarify wine and beer. Tannic acid was once used to denature industrial alcohol. Tannins are used to tan leather, which used to involve soaking hides in a solution derived from tree bark, thereby allowing tannic acid to permeate the skin. The result was that it became supple and immune from bacterial attack.

tartrate: the salts of tartaric acid, $C_4H_6O_4$, are called tartrates. Potassium tartrate is called cream of tartar. The acid itself consists of colourless, transparent crystals that are soluble in water. Tartaric acid occurs naturally in wine lees, and it can be made synthetically in other ways. It is a traditional ingredient in baking powder, and is used as the solid acid in fizzy indigestion and headache tablets. When these are put into water, the acid dissolves and reacts with sodium bicarbonate to release carbon dioxide gas. Tartaric acid occurs in two molecular forms which differ only in that

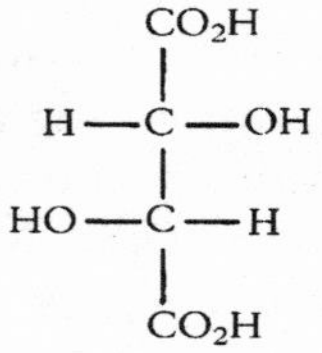

L-tartaric acid

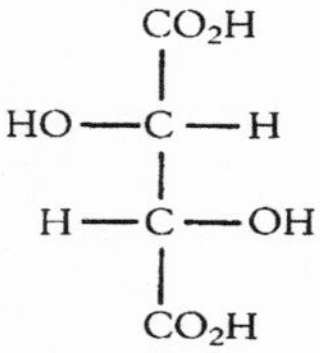

D-tartaric acid

one is the mirror image of the other. The D–form is the natural one produced in the wine industry.

TCDD: the abbreviation for 2,3,7,8-tetra-chlorodibenzodioxin, in which there are four chlorine atoms attached to the dibenzo–dioxin molecule (see **dioxin**).

tetrafluoroethylene: C_2F_4, also known as TFE and perfluoroethylene, and more correctly as tetrafluoroethene. It is a gas, with boiling point $-78\,°C$, and is the starting material from which polytetrafluoroethylene is made (see **PTFE**).

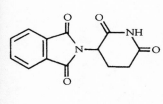

thalidomide: $C_{13}H_{10}N_2O_4$. There are two forms of thalidomide which have identical chemical formulae, and which consist of exactly the same groups of interlinked rings of atoms. The difference is that one molecule is the mirror image of the other, and they are labelled S (from the Latin *sinister*) or R (*rectus*). R-thalidomide is safe, but if the S form is taken on certain days of a pregnancy, it will interfere with the replicating DNA and deform the foetus. As it was originally manufactured, thalidomide consisted of both R and S molecules in equal amounts.

titanium dioxide: TiO_2, a brilliant white solid that is produced on a large scale and used as the chief pigment in paints because of its covering power. It occurs naturally as the minerals anatase and rutile, but the TiO_2 used commercially has to be pure and is obtained, for example, by converting the titanium to its chloride $TiCl_4$ and then reacting this with water. TiO_2 is non–combustible and non-toxic to humans and is used in a variety of ways as a 'filler' (inert bulking agent) for paper and plastics. It is also used in cosmetics, glassware, ceramics, and printing inks.

toluene: C_7H_8, an aromatic hydrocarbon which consists of a methyl group attached to a benzene ring, and sometimes called methylbenzene. Toluene is a colourless liquid which boils at 111 °C and has a density of 0.867 grams per millilitre. As a solvent it has all the dissolving power of benzene, and is used in industry and research instead of benzene.

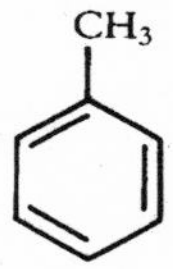

2,4,6-trichlorophenol: $C_6H_3Cl_3O$. The numbers refer to the positions of the three chlorine atoms attached to the benzene ring. It is a yellow solid which melts at 61 °C and is produced by the reaction of chlorine gas and phenol. It is a plant fungicide, herbicide, and wood preservative, although it is now little used because it is contaminated with dioxins.

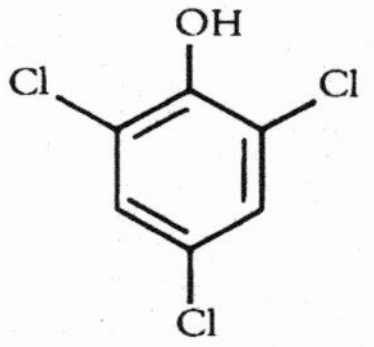

2,4,5-T: $C_8H_5Cl_3O_3$, also known as 2,4,5-trichlorophenoxyacetic acid, where the numbers refer to the positions of the three chlorine atoms attached to the benzene ring. It is a pale brown solid which melts at 151 °C. 2,4,5-T has been widely used as a plant hormone, herbicide and defoliant, although it is contaminated with dioxins. See also **2,4-D**.

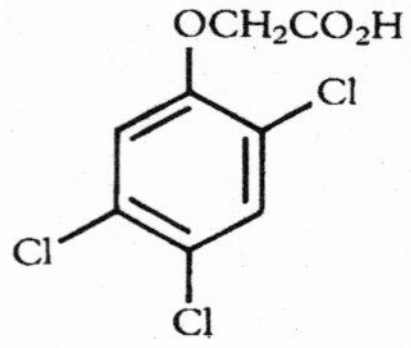

triglycerides: these are any naturally occurring esters of glycerol to which are attached long chain fatty acids. One such fatty acid is stearic acid with the formula $C_{17}H_{35}CO_2H$, which has a long chain of 17 carbon atoms that is saturated. If three of these are attached to **glycerol** it forms tristearin, which is a major component of beef fat and cocoa butter. The fatty acid oleic acid has the formula $C_{17}H_{33}CO_2H$, and its chain of 17 carbon atoms has a double bond about half way along. Oleic acid is an unsaturated acid, and is the principal acid component of olive oil.

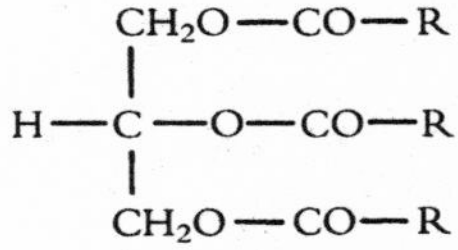

R = long chain of fatty acid

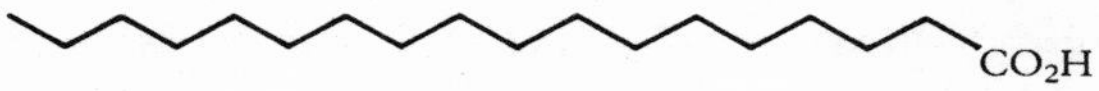

stearic acid

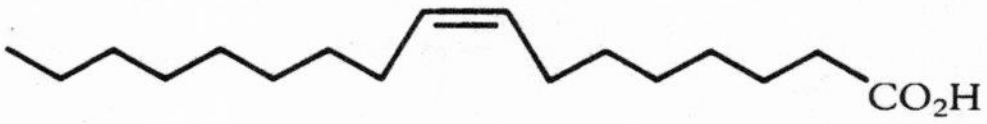

oleic acid

trinitrotoluene (TNT): the basic molecule which can give rise to musk-like fragrances such as *tert*-butyl-trinitrotoluene and musk ketone, $C_{14}H_{18}N_2O_5$, which have warm, sweet, erogenous musk-like odours.

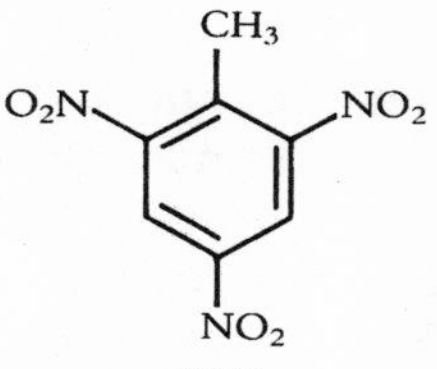

TNT

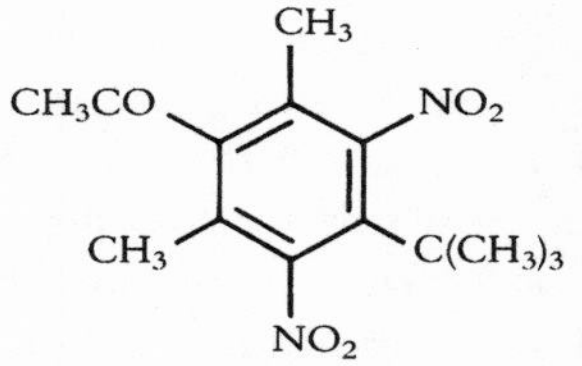

musk ketone

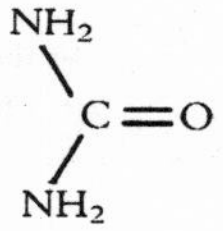

urea: CH_4N_2O, a stable, white, crystalline solid that is one of the waste products of nitrogen metabolism in animals and one of the main components of urine. It makes a good fertiliser and can be manufactured from ammonia and carbon dioxide.

Valium: the tradename for **diazepam**.

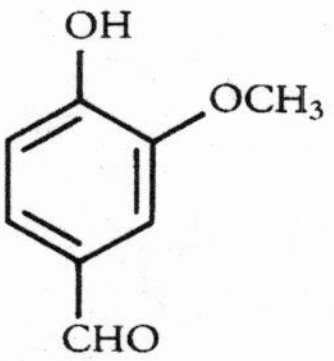

vanilla: $C_8H_8O_3$, also known as vanillin, a creamy white crystalline material which has the chemical name 3-methoxy-4-hydroxy-benzaldehyde. It is manufactured from the lignin waste of the wood pulp industry. It occurs naturally in the vanilla pod, in balsams and in benzoins. Vanillin is used in perfumes for its intense, sweet, creamy odour. It is also used in food flavours and to make L-dopa (L-dihydroxyphenylalanine), the amino acid used in treating Parkinson's disease.

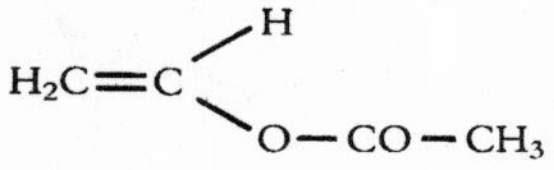

vinyl acetate: $C_4H_6O_2$, a derivative of ethylene and the basic building block of poly(vinyl acetate) or PVA.

vinyl chloride: C_2H_3Cl, also known as vinyl chloride monomer, and abbreviated to VCM, but more correctly known as chloroethene. This is the starting material from which **PVC** is made.

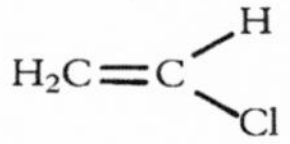

$$H_2C{=}C\diagup^H_{\diagdown Cl}$$

xenon: element, atomic number 54, atomic mass 131, symbol Xe. This element is very unreactive and in the air it is found only as a gas composed of single atoms. It condenses to a liquid at $-107\,°C$. Xenon is rare and accounts for only 0.09 ppm of the atmosphere, although this in total comes to about 2 billion tonnes (2×10^9). A little is extracted each year from liquid air and used for chemical research. Although xenon is completely inert towards most other materials, it can be made to react with fluorine gas (F_2) to form solid compounds such as XeF_2, XeF_4 and XeF_6. These are highly reactive.

xylitol: $C_5H_{12}O_5$, a white, sweet-tasting, crystalline powder, stable in air, soluble in alcohol and very soluble in water. A litre of water will dissolve 1.6 kg of xylitol.

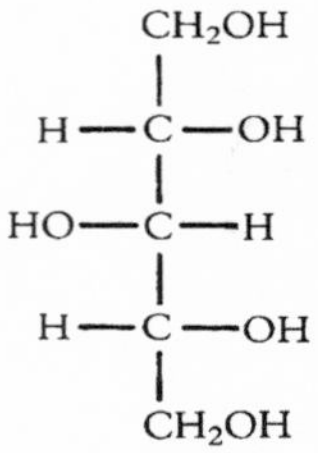

$$
\begin{array}{c}
CH_2OH \\
| \\
H{-}C{-}OH \\
| \\
HO{-}C{-}H \\
| \\
H{-}C{-}OH \\
| \\
CH_2OH
\end{array}
$$

xylose: $C_5H_{10}O_5$, also known as wood sugar or birch sugar, and more correctly as α-D-pyranose. It is a white crystalline powder which is stable in air and soluble in water.

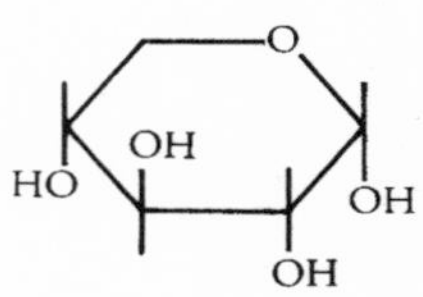

yohimbine: $C_{21}H_{26}O_3N_2$, also known as aphrodine, corynine and quebrachine. It is a member of the group of chemicals known as alkaloids, and forms needle-like crystals when pure which melt at $234\,°C$. It is extracted from the bark of the *Corynanthe yohimbe* tree which grows in Cameroon.

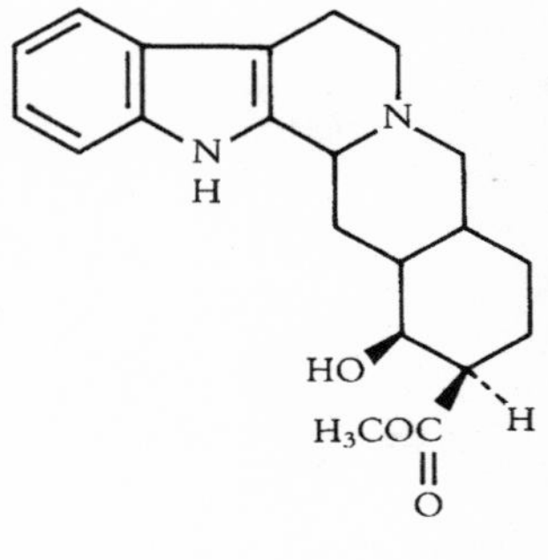

BIBLIOGRAPHY

This bibliography contains the main sources of information I have used in writing the book. Under each chapter heading I have included a few titles that give good general coverage and which are written for people without specialist chemical training. The other references do assume a specialist knowledge in the area, and these are sources that you would probably only find in a reference or college library.

Chapter 1 Perfumes

General reading

A. Birchall, 'A whiff of happiness', *New Scientist*, p. 44, 25 August 1990.

P. M. Muller and D. Lamparsky, eds, *Perfumes: Art, Science and Technology*, Elsevier, Amsterdam, 1991.

D. M. Stoddart, *The Scented Ape*, Cambridge University Press, 1990.

S. Van Toller and G. H. Dodd, eds, *Perfumery: the Psychology and Biology of Fragrance*, Chapman and Hall, London, 1991.

Specialist reading

J. Ayres, 'Uses of fine chemicals in flavours and fragrances', *Chemistry & Industry*, p. 575, 19 September 1988.

R. J. Kutsky, *Handbook of Vitamins, Minerals and Hormones*, 2nd edition, Van Nostrand Reinhold, New York, 1981.

J. Müller *et al.*, *The H&R Book of Perfume: Understanding Fragrance, Origin, History, Development and Guide to Fragrance Ingredients*, Glöss Verlag, Hamburg, 1992.

C. S. Sell, 'Organic chemistry in the perfume industry', *Chemistry in Britain*, p. 791, August 1988.

C. S. Sell, 'The chemistry of ambergris', *Chemistry in Britain*, p. 516, August 1990.

Fragrance Guide, Feminine Notes and Masculine Notes, Glöss Verlag, Hamburg, 1991.

Chapter 2 Sweetness and light

General reading

H. Hobhouse, *Seeds of Change: Five Plants that Transformed Mankind*, Sidgwick & Jackson, London, 1985. This has a chapter on the history of sugar.

R. Tannahill, *Food in History*, revised edition, Penguin, 1988.

H. D. Belitz and W. Grosch, *Food Chemistry*, 2nd edition, translated by D. Hadziyev, Springer Verlag, Berlin, 1987.

J. Emsley and L. Hough, 'Sweeter by far?' *New Scientist*, p. 48, 19 June 1986.

J. Yudkin, *Pure White and Deadly*, updated edition, Viking, London, 1986.

Specialist reading

L. O'Brien Nabors and R. C. Gelardi, eds, *Alternative Sweeteners*, Marcel Dekker, New York, 1986.

L. Oxdale, 'The sweet taste of success', *Chemistry in Britain*, March 1991.

T. H. Grenby, 'Prospects for sugar substitutes', *Chemistry in Britain*, April 1991.

H. Schiweck, K. Rapp and M. Vogel, 'Utilization of sucrose as an industrial chemical, state of the art and future applications', *Chemistry & Industry*, 4 April 1988.

B. Crammer and R. Ikan, 'Sweet glycosides from the stevia plant', *Chemistry in Britain*, October 1986.

Developments in Sweeteners, Elsevier Applied Science, London, 1979 – An on-going series of volumes edited by various people, and containing articles written by authorities in each area.

Chapter 3 Alcohol

General reading

T. Parratt, *Name Your Poison, A Guide to Additives in Drinks*, Robert Hale, London, 1990.

E. Maury, *Your Good Health, The Medical Benefits of Wine Drinking*, Souvenir Press, London, 1992.

J. Postgate, *Microbes and Man*, 3rd edition, Penguin, London, 1992.

Specialist reading

H.-D. Belitz and W. Grosch, *Food Chemistry*, 2nd edition, translated by D. Hadziyev, Chapter 20, Springer Verlag, Berlin, 1987.

R. Boddey, 'Green energy from sugar cane', *Chemistry and Industry*, 17 May 1993, p. 355.

F. A. Lowenheim and M. K. Moran, *Faith, Keyes, and Clark's Industrial Chemicals*, 4th edition, Wiley-Interscience, New York, 1975.

The Essential Chemical Industry, Chemical Industry Education Centre, University of York, 1989.

R. Muller, 'Flavour in low-alcohol beers', *Chemistry Review*, September 1993.

G. Winger, F. G. Hofmann and J. H. Woods, *A Handbook on Drug and Alcohol Abuse, Biomedical Aspects*, 3rd edition, Oxford University Press, 1992.

Chapter 4 Cholesterol, fats and fibre

General reading

A. E. Bender, *Health or Hoax?* Sphere, London, 1986.

S. Bingham, *The Everyman Companion to Food and Nutrition*, J. M. Dent, London, 1987.

Robert, L. Ory, *Grandma called it Roughage, Fiber Facts and Fallacies*, American Chemical Society, Washington DC, 1991.

A. Eyton, *The F-plan Diet*, Penguin, 1982.

R. E. Kowalski, *The 8-Week-Cholesterol Cure*, Bantam, Sydney, 1988.

C. A. Rinzler, *Food Facts and What They Mean*, Bloomsbury, London, 1987.

Specialist reading

H.-D. Belitz and W. Grosch, *Food Chemistry*, Springer Verlag, Berlin, 1987, Chapter 14.

T. P. Coultate, *Food, the Chemistry of its Components*, 2nd edition, Royal Society of Chemistry, London, 1988.

J. Davies and J. Dickerson, *Nutrient Content of Food Portions*, Royal Society of Chemistry, London, 1991.

R. Passmore and M. A. Eastwood, *Davidson and Passmore, Human Nutrition and Dietetics*, 8th edition, Churchill Livingstone, Edinburgh, 1986.

A. A. Paul and D. A. T. Southgate, *McCance and Widdowson's The Composition of Foods*, 4th edition, Her Majesty's Stationery Office, London, 1988.

Chapter 5   Painkillers and painful decisions

General reading

Living with Risk: the British Medical Association Guide, John Wiley, Chichester, 1987.

B. Selinger, *Chemistry in the Market Place*, 4th edition, Harcourt Brace Jovanovich, Sydney, 1988, Chapter 9, pp. 300–344.

M. Weatherall, *In Search of a Cure, a History of Pharmaceutical Discovery*, Oxford University Press, 1990.

Specialist reading

S. Adams, 'The discovery of brufen (ibuprofen)', *Chemistry in Britain*, December 1987.

A. Albert, *Xenobiosis, Food, Drugs and Poisons in the Human Body*, Chapman and Hall, London 1987.

S. Stinson, *Better Understanding of Arthritis Leading to New Drugs to Treat it*, Product Report, *Chemical & Engineering News*, 16 October 1989.

J. A. Timbrell, *Introduction to Toxicology*, Taylor & Francis, London, 1989.

Chapter 6 PVC

General reading

H. G. Elias, *Mega Molecules*, Springer Verlag, Berlin, 1985.

N. R. Kamsvåg and J. Baldwin, eds, *PVC and the Environment*, Norsk Hydro, 1992. This is a comprehensive 220-page book for the general reader. It gives a pro-industry view, but much was written by academic experts in the field. The book also tries to carry out a full cradle-to-grave analysis of PVC, including an environmental and health audit. Norsk Hydro also publishes a 24-page colour brochure *Facts about PVC*, which is available from their Petrochemical Division, P. O. Box 2594, Solli, 0203 Oslo 2, Norway.

Specialist reading

C. A. Heaton, ed., *The Chemical Industry*, Blackie, Glasgow, 1986.

D. Houndshell and J. Kelly Smith, *Science and Corporate Strategy, DuPont R&D 1902–80*, Cambridge University Press, 1988.

C. Kennedy, *ICI, the Company that Changed our Lives*, Hutchinson, London, 1986.

R. J. Lewis Sr, *Carcinogenically Active Chemicals*, Van Nostrand Reinhold, New York, 1991.

P. J. T. Morris, *Polymer Pioneers*, The Center for History of Chemistry, publication No. 5, Philadelphia, 1986.

M. P. Stevens, 'Polymer additives', *Journal of Chemical Education*, p. 444, Volume 70, 1993.

W. Tötsch and J. Gaensslen, *Polyvinylchloride. Environmental Aspects of a Common Plastic*, Elsevier, 1992.

Chapter 7 Dioxins

General reading

H. D. Crone, *Chemicals & Society*, Cambridge University Press, 1986.

M. A. Ottobone, *The Dose Makes the Poison*, 2nd edition, Van Nostrand Reinhold, New York, 1991.

Specialist reading

The journal *The Science of the Total Environment* published a special issue, Volume 104, entitled 'Dioxins. Chemistry and Health Effects' in 1991. Some of the key papers are included below.

B. N. Ames, 'Natural carcinogens and dioxin', *The Science of the Total Environment*, 1991. Volume 104, pp. 159–166.

T. Colborn and C. Clement, eds, *Chemically-Induced Alterations in Sexual and Functional Development: The Wildlife/Human Connection*, in the series *Advances in Modern Environmental Toxicology*, Volume XXI, Princeton Scientific Publishing, New Jersey, 1992.

Dioxins in the Environment, Pollution Paper No. 27, Department of the Environment, Her Majesty's Stationery Office, 1989.

G. H. Eduljee, 'Dioxins in the environment', *Chemistry in Britain*, December 1988.

H. Fiedler, O. Hutzinger and C. W. Timms, 'Dioxins: sources of environmental load and human exposure', *Toxicological and Environmental Chemistry*, Volume 29, 1990, pp. 157–234.

M. A. Gallo, R. J. Scheuplein and K. A. Van der Heijden, eds, *Banbury Report 35: Biological Basis for Risk Assessment of Dioxins and Related Compounds*, Cold Spring Harbor Laboratory, 1991.

M. Gough, 'Human health effects: what the data indicate', *The Science of the Total Environment*, 1991. Volume 104, pp. 129–158.

G. W. Gribble, 'Naturally occurring organohalogen compounds—a survey', *Journal of Natural Products*, 1992, Volume 55, pp. 1353–1395.

D. Hanson, 'Dioxin toxicity: new studies . . .', *Chemical & Engineering News*, 12 August 1991.

S. J. Harrad and K. Jones, 'Dioxins at large', *Chemistry in Britain*, December 1992.

C. A. Heaton, ed., *The Chemical Industry*, Blackie, Glasgow and London, 1986.

O. P. Kharbanda and E. A. Stallworthy, *Safety in the chemical industry*, Chapter 11, Heinemann, London, 1988.

L. O. Kjeller, K. C. Jones, A. E. Johnston and C. Rappe, 'Increases in the polychlorinated dibenzo-p-dioxin and -furan content of soils and vegetation since the 1840s; *Environmental Science and Technology*, Volume **25**, pp. 1619–1627, 1991.

A. Manz, J. Berger, J. H. Dwyer, D. Flesch-Janys, S. Nagel and H. Waltsgott, 'Cancer mortality among workers in chemical plant contaminated with dioxin', *The Lancet*, Volume 338, p. 959, 1991.

C. Rappe and H. R. Buser, 'Chemical and physical properties, analytical methods, sources and environmental levels of halogenated dibenzo-dioxins and dibenzo-furans,' *Halogenated Biphenyls, Terphenyls, Naphthalenes, Dibenzodioxins and Related Products* (eds Kimbrough and Jensen), Chapter 3, Elsevier Science Publishers, 1989.

Chapter 8   Nitrate

General reading

T. M. Addiscott, 'Farmers, fertilizers and the nitrate flood,' *New Scientist*, 8 October 1988, p. 50.

T. M. Addiscott, A. P. Whitmore and D. S. Powlson, *Farming, Fertilizers and the Nitrate Problem*, C·A·B International, Wallingford (UK), 1991.

O. C. Bøckman, O. Kaarstad, O. H. Lie and I. Richards, *Agriculture and Fertilizers*, Norsk Hydro, Oslo (Norway), 1990.

Specialist reading

C. Glidewell, 'The nitrate/nitrite controversy', *Chemistry in Britain*, February 1990, p. 137.

K. W. T. Goulding and P. R. Poulton, 'Unwanted nitrate', *Chemistry in Britain*, December 1992, p. 1100.

Ministry of Agriculture, Fisheries and Food (MAFF) booklet, *Solving the Nitrate Problem*, 1993. Available free from MAFF Publications, London SE99 7TP.

J. I. Prosser, ed., *Nitrification*, IRL Press, Oxford, 1986.

Chapter 9   Carbon dioxide, CO_2

General reading

R. C. Balling Jr., *The Heated Debate*, Pacific Research Institute for Public Policy, San Francisco, 1992.

R. Bate and J. Morris, *Apocalypse or Hot Air?*, Institute of Economic Affairs, London, 1994.

N. Fell and P. Liss, 'Can algae cool the planet?' *New Scientist*, 21 August 1993.

J. Gribbin, *Hothouse Earth: the Greenhouse Effect and Gaia*, Black Swan, London, 1989.

J. Leggett (ed.), *Global Warming, the Greenpeace Report*, Oxford University Press, 1990.

J. Mason, 'The greenhouse effect and global warming', Royal Institution lecture, London, November 1990. Available from British Coal, Coal Research Establishment, Stoke Orchard, Cheltenham, Gloucestershire GL52 4RZ, UK.

F. Pearce, 'The high cost of carbon dioxide', *New Scientist*, 17 July 1993.

Specialist reading

T. E. Graedel, D. T. Hawkins and L. D. Claxton, *Atmospheric Chemical Compounds, Sources, Occurrence and Bioassay*, Academic Press, Orlando, Florida, 1986.

The Hadley Centre Transient Climate Change Experiment, August 1992. Report from the Hadley Centre for Climate Prediction, Meteorological Office, London Road, Bracknell, Berks RG12 2SY, UK.

J. T. Houghton, G. T. Jenkins, and J. J. Ephraums (eds), *Scientific Assessment of Climate Change*, WMO/UNEP Intergovernmental Panel on Climate Change, Geneva, 1990, also published by Cambridge University Press. A summary booklet is available as the *1990 IPCC Scientific Assessment of Climate Change*, Submission from Working Group 1 to the Intergovernmental Panel on Climate Change (IPCC), World Meteorological Organization, United Nations Environment Programme, 1990.

1992 IPCC Supplement Scientific Assessment of Climate Change, Submission from Working Group 1 to the Intergovernmental Panel on Climate Change (IPCC), World Meteorological Organization, United Nations Environment Programme, 1992.

Z. Jaworowski, T. V. Segalstad and N. Ono, 'Do glaciers tell a true atmospheric CO_2 story?' *The Science of the Total Environment*, Volume 114, pp. 227–284, 1992.

Z. Jaworowski, T. V. Segalstad and V. Hisdsal, *Atmospheric CO_2 and Global Warming: a Critical Review*, 2nd revised edition, Norsk Polar Institutt, Oslo, 1992.

R. H. Nielsen, 'Chill warnings from Greenland', *New Scientist*, 28 August 1993. Based on two articles in *Nature*, Volume 364, pp. 203 and 218, 1993.

P. Warneck, *Chemistry of the Natural Atmosphere*, Academic Press, San Diego, California, 1988.

R. P. Wayne, *Chemistry of Atmospheres*, 2nd edition, Clarendon Press, Oxford, 1991.

Chapter 10   Take care

General reading

If you want to read further about the role of chemistry and chemicals in every day life, then the following books are highly recommended.

P. W. Atkins, *Molecules*, Scientific American Library, New York, 1987.

H. D. Crone, *Chemicals and Society*, Cambridge University Press, 1986.

Living with Risk: the British Medical Association Guide, John Wiley, Chichester, 1987.

J. Lenihan, *The Crumbs of Creation*, Adam Hilger, Bristol, 1988.

M. A. Ottoboni, *The Dose Makes the Poison*, 2nd edition, Van Nostrand Reinhold, New York, 1991.

J. V. Rodricks, *Calculated Risks, Understanding the Toxicity and Human Health Risks of Chemicals in our Environment*, Cambridge University Press, 1992.

B. Selinger, *Chemistry in the Market Place*, 4th edition, Harcourt Brace Jovanovich, Sydney, 1989.

Appendix   Chemical Data

General reading

A. Nickon and E. F. Silversmith, *Organic Chemistry, The Name Game, Modern Coined Terms and Their Origins*, Permanon Press, New York, 1987.

Specialist reading

H.-D. Belitz and W. Grosch, *Food Chemistry*, Springer Verlag, Berlin, 1987.

S. Budavari, ed., *The Merck Index*, 11th edition, Merck, Rahway, New Jersey, 1989.

J. Emsley, *The Elements*, 2nd edition, Oxford University Press, 1992.

G. G. Hawley, *The Condensed Chemical Dictionary*, 10th edition, Van Nostrand Reinhold Company, New York, 1981.

R. C. Weast, editor-in-chief, *CRC Handbook of Chemistry and Physics*, 70th edition, CRC Press, Boca Raton, Florida, 1990.

Dictionary of Organic Compounds, 5th edition, Chapman and Hall, London and New York, 1982, plus supplements.

Dictionary of Inorganic Compounds, Chapman and Hall, London and New York, 1992, plus supplement.

INDEX

Chemical compounds are listed alphabetically, but not including single letter Greek and Roman prefixes.